PULSED ELECTRIC FIELDS IN FOOD PROCESSING

FOOD PRESERVATION TECHNOLOGY SERIES

Series Editor: Gustavo V. Barbosa-Cánovas

Innovations in Food Processing
    Editors: Gustavo V. Barbosa-Cánovas
              Grahame W. Gould

Trends in Food Engineering
    Editors: Jorge E. Lozano
              Cristina Añón
              Efrén Parada-Arias
              Gustavo V. Barbosa-Cánovas

Pulsed Electric Fields in Food Processing: Fundamental Aspects and Applications
    Editors: Gustavo V. Barbosa-Cánovas
              Q. Howard Zhang

Osmotic Dehydration and Vacuum Impregnation: Applications in Food Industries
    Editors: Pedro Fito
              Amparo Chiralt
              Jose M. Barat
              Walter E. L. Spiess
              Diana Behsnilian

Further information on current and forthcoming titles,
including Tables of Contents, is available at **www.techpub.com**

FOOD PRESERVATION TECHNOLOGY SERIES

Pulsed Electric Fields in Food Processing

Fundamental Aspects and Applications

EDITED BY

Gustavo V. Barbosa-Cánovas
Washington State University

Q. Howard Zhang
The Ohio State University

ASSOCIATE EDITOR

Gipsy Tabilo-Munizaga
Washington State University

Routledge
Taylor & Francis Group

LONDON AND NEW YORK

Pulsed Electric Fields in Food Processing

First published 2001 by Technomic Publishing Company, Inc.

Published 2018 by Routledge
2 Park Square, Milton Park, Abingdon, Oxon OX14 4RN
52 Vanderbilt Avenue, New York, NY 10017

Routledge is an imprint of the Taylor & Francis Group, an informa business

First issued in paperback 2019

Main entry under title:
    Food Preservation Technology Series: Pulsed Electric Fields in Food Processing—
    Fundamental Aspects and Applications

Bibliography: p.
Includes index p. 265

Library of Congress Catalog Card No. 00-111666

ISBN 13: 978-0-367-45533-0 (pbk)
ISBN 13: 978-1-56676-783-5 (hbk)

To our families

Table of Contents

Series Preface

T HIS is the third book released as part of the Technomic Food Preservation Technology Series, after *Innovations in Food Processing* and *Trends in Food Engineering*. It deals with the preservation of foods by Pulsed Electric Fields, one of the most promising alternative technologies to process foods. This edited book includes authors from industry and academia located in different parts of the world. The editors are fully engaged with this technology and they are the leaders, in their respective universities, in developing PEF technology from laboratory scale to industrial size.

The topic of Pulsed Electric Fields fits very nicely with the goals of the Series, identified sometime ago, which are:

- publish books on topics of current interest
- cover the selected topic from fundamentals to applications
- collect in a single volume the opinions of the most authoritative food scientists and technologists in the subject
- present a clear picture of the impact of the selected topic on the world food domain
- address, in a comprehensive manner, key issues of a given technology, such as food safety, regulation, inactivation of microorganisms and enzymes, engineering aspects, sensory, quality, shelf-life, and market opportunities

This book, in my opinion, will set the stage for the future development of the use of PEF in the food sector. It will be followed very shortly by other timely and important titles, including *Osmotic Dehydration, Engineering and Food for the 21st Century, Food Science and Food Biotechnology*, and *Alternative Food Processing Technologies*.

This Series is growing rapidly and it is making a positive impact toward the understanding and development of sound strategies to process and characterize the foods of today and tomorrow.

GUSTAVO V. BARBOSA-CÁNOVAS
Series Editor

Preface

Pulsed electric fields (PEF) is one of the nonthermal processing approaches that is receiving considerable attention from scientists, government, and the food industry as a potential technique to be fully adopted to process foods at the industrial level. PEF presents a number of advantages including minimal changes to fresh foods, and inactivation of a wide range of microorganisms and enzymes. It also offers the opportunity to develop new food products not feasible through conventional thermal processing.

PEF is under scrutiny by R&D groups around the world. Typical endeavors for these groups are microbial inactivation, tissue response to electric fields, enzyme inactivation, engineering aspects, modeling, and scale-up studies. This technology is constantly evolving as demonstrated by the many technical contributions on this subject presented throughout the world at relevant scientific gatherings, the availability of funding allocated to further explore this technology, the interest in knowing more about PEF by regulatory agencies, and the production trials currently underway at several key food manufacturing companies and food processing equipment suppliers.

This book includes 17 research contributions written by scientists working in this field for a good number of years. Chapter 1 presents a very comprehensive review of PEF, both the microbial and engineering aspects. Chapter 2 is dedicated to the study of flow patterns in a PEF continuous system, in order to identify optimal operational flow rates. It is followed by an analysis of key physical properties of this technology, including measured values for all of them in specific foods. Chapter 4 reviews the key studies conducted so far in the inactivation of enzymes by PEF. The next five chapters cover, in great detail, thorough studies on the inactivation of very specific enzymes in selected media such as alkaline phosphatase in milk, extracellular proteases in casein, and endoproteases in faba bean protein concentrate hydrolysate. Chapters 10 and 11 deal with PEF inactivation of three important microorganisms, *Listeria innocua*, *Pseudomonas fluorescens*, and *Bacillus subtilis* spores. Chapters 12 and 13 are also on microbial inactivation, but they include a comparison in efficacy with high hydrostatic pressure (HHP), another promising nonthermal technology. The microorganisms under consideration are *Saccharomyces cerevisiae* and *Pseudomonas fluorescens*. Chapter 14 covers the reformulation of two food products, cheese sauce and salsa, to make them more feasible for processing by PEF. Chapter 15 includes a thorough comparison on the electrophoretic patterns of liquid whole eggs processed by PEF, HHP, and thermal processing. Chapter 16 deals with a comparison of these three technologies in the processing of raw apple juice in terms of shelf stability, sensory analysis, and volatile flavor profile. Finally, Chapter 17 includes some considerations, from the industrial point of view, of PEF food processing in relation to safety assurance.

We feel that these original contributions will significantly help in a better understanding of the use of pulsed electric field technology to process food, and we sincerely hope this book will promote additional interest in pulsed electric field technology research, development, and implementation.

GUSTAVO V. BARBOSA-CÁNOVAS
Q. HOWARD ZHANG

Acknowledgements

THE editors want to recognize all the help received from the associate editor, Gipsy Tabilo-Munizaga (Washington State University), the authors, and the reviewers in making this book a reality. We also thank Jeannie Andersen and Dora Rollins, both with Washington State University, for their editorial comments.

List of Contributors

GUSTAVO V. BARBOSA-
  CÁNOVAS
Department of Biological Systems
  Engineering
Washington State University
Pullman, WA 99164-6120 U.S.A.

EMMA BARKSTROM
Department of Biological Systems
  Engineering
Washington State University
Pullman, WA 99164-6120 U.S.A.

ANA S. BERMÚDEZ
Chemistry Department
National University of Colombia
Santa Fé de Bogotá
D.C. 5997, Colombia

TERRY BOYLSTON
Department of Food Science and
  Human Nutrition
Washington State University
Pullman, WA 99164-6376 U.S.A.

ARMANDO J. CASTRO
Department of Food Science and
  Human Nutrition
Washington State University
Pullman, WA 99164-6376 U.S.A.

FU J. CHANG
Department of Biological Systems
  Engineering
Washington State University
Pullman, WA 99164-6120 U.S.A.

GRADY W. CHISM
Department of Food Science and
  Technology
The Ohio State University
2121 Fyffe Road
Columbus, OH 43210 U.S.A.

P. MICHAEL DAVIDSON
Food Research Center
University of Idaho
Moscow, ID 83843 U.S.A.

A. KEITH DUNKER
Biochemistry and Chemistry
  Department
Washington State University
Pullman, WA 99164 U.S.A.

JOSEPH DUNN
R&D Automatic Liquid
  Packaging, Inc.
2200 West Lakeshore Drive
Woodstock, IL 60098 U.S.A.

SANTIAGO ESPLUGAS
University of Barcelona
Department of Chemical
  Engineering
08028 Barcelona, Spain

JUAN J. FERNÁNDEZ-MOLINA
Department of Biological Systems
  Engineering
Washington State University
Pullman, WA 99164-6120 U.S.A.

M. MARCELA GÓNGORA-NIETO
Department of Biological Systems
  Engineering
Washington State University
Pullman, WA 99164-6120 U.S.A.

W. JAMES HARPER
Department of Food Science and
  Technology
The Ohio State University
2121 Fyffe Road
Columbus, OH 43210 U.S.A.

STEVE L. HARRISON
Department of Food Science and
  Human Nutrition
Washington State University
Pullman, WA 99164-6376 U.S.A.

Z. TONY JIN
Department of Food Science and
  Technology
The Ohio State University
2121 Fyffe Road
Columbus, OH 43210 U.S.A.

HUUB L. M. LELIEVELD
Unilever Research Vlaardingen
Vlaardingen, The Netherlands

ALAIN E. LEÓN
Unilever Health Institute
P.O. Box 114
3130 AC
Vlaardingen, The Netherlands

LLOYD LUEDECKE
Department of Food Science and
  Human Nutrition
Washington State University
Pullman, WA 99164-6376 U.S.A.

LI MA
Department of Biological Systems
  Engineering
Washington State University
Pullman, WA 99164-6120 U.S.A.

OLGA MARTÍN-BELLOSO
Department of Food Technology
University of Lleida
Spain

RAFAEL PAGÁN
University of Zaragoza
Faculty of Veterinary
50013 Zaragoza
Spain

LILIAN A. PALOMEQUE
Chemistry Department
National University of Colombia
Santa Fé de Bogotá
D.C. 5997, Colombia

JOSEPH R. POWERS
Department of Food Science and
  Human Nutrition
Washington State University
Pullman, WA 99164-6376 U.S.A.

PHILIPPE RIQUET
Department of Biological Systems
  Engineering
Washington State University
Pullman, WA 99164-6120 U.S.A.

KATHRYN T. RUHLMAN
Department of Food Science and
  Technology
The Ohio State University
2121 Fyffe Road
Columbus, OH 43210 U.S.A.

SUDHIR K. SASTRY
Department of Food Science and
  Technology
The Ohio State University
2121 Fyffe Road
Columbus, OH 43210 U.S.A.

LYDIE SEIGNOUR
Department of Biological Systems
  Engineering
Washington State University
Pullman, WA 99164-6120 U.S.A.

YIPING SU
Department of Food Science and
  Technology
The Ohio State University
2121 Fyffe Road
Columbus, OH 43210 U.S.A.

BARRY G. SWANSON
Department of Food Science and
  Human Nutrition
Washington State University
Pullman, WA 99164-6376 U.S.A.

PONTUS TORSTENSSON
Department of Biological Systems
  Engineering
Washington State University
Pullman, WA 99164-6120 U.S.A.

LAURA TUHELA
Department of Food Science and
  Technology
The Ohio State University
2121 Fyffe Road
Columbus, OH 43210 U.S.A.

HUMBERTO VEGA-MERCADO
Department of Biological Systems
  Engineering
Washington State University
Pullman, WA 99164-6120 U.S.A.

PATRICK C. WOUTERS
Unilever Research Vlaardingen
Vlaardingen, The Netherlands

HYE W. YEOM
Department of Food Science and
  Technology
The Ohio State University
2121 Fyffe Road
Columbus, OH 43210 U.S.A.

AHMED E. YOUSEF
Department of Food Science and
  Technology
The Ohio State University
2121 Fyffe Road
Columbus, OH 43210 U.S.A.

Q. HOWARD ZHANG
Department of Food Science and
  Technology
The Ohio State University
2121 Fyffe Road
Columbus, OH 43210 U.S.A.

CHAPTER 1

Pulsed Electric Field Processing: An Overview

J. DUNN

INTRODUCTION

T HE roots of pulsed electric field processing can be traced to Germany. Doevenspeck's patent (Dovenspeck 1960), dating to the 1960s, describes a variety of pulsed electrical field (PEF) equipment and methods ranging from PEF processing of sausage to specific electronic embodiments (Figure 1.1). This inventive fellow remained active for many years and collaborated on PEF development with later German investigators. At the time of his death, Doevenspeck and Sitzmann (the inventor of the Elcrack and Elsteril PEF methods during his tenure at Krupp Machinentechnick in Hannover) (Sitzmann, 1990, 1995) were collaborating on a PEF program funded by the German government. The contributions of Doevenspeck over approximately 50 years should not be underrated; he could rightfully be called the father of PEF processing.

In the 1960s, Sale and Hamilton (Sale and Hamilton, 1967; Hamilton and Sale, 1967), working at Unilever, published a short series of insightful papers that, even today, provide valuable reference. Their studies made several important observations, including early mechanistic studies on cellular repair after PEF treatment.

Many groups have rediscovered PEF while investigating other phenomena. Examples include studies using impedance cell counting methods (popularized and commercialized by Wallace Coulter, founder of Coulter Electronics) and experiments conducted to study rapid temperature excursions (temperature jump experiments) produced via electrical pulsation. In the former, studies of the relationship between gap voltage and cell impedance showed an anomalous sudden change in slope (toward greater conductivity and lower cell

1

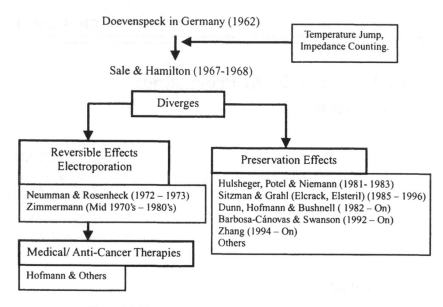

Figure 1.1 Pioneers in the development of pulsed electric field.

impedance/resistance) above certain gap field strengths (hence the term "dielectric breakdown"). In the latter, some of the results attributed to rapid temperature jump seem, in hindsight, to be related to electric field effects.

PEF development then diverged (Figure 1.2) into two different, though related, areas of endeavor: (1) reversible electroporation, or PEF performed under conditions promoting electroporation and cell survival; and (2) PEF for microorganism inactivation and food preservation. Microbiological inactivation and food preservation effects relied on multiple high strength electric field pulses (greater than ~18 kV/cm) of relatively short duration (sub-μs to

Equipment & Treatments

PEF	**Reversible Electroporation**
❖ Different chamber & equipment design ❖ Relatively high e-field (>18 kV/cm) ❖ Short duration pulses (<5 μs)	❖ Relatively low e-field (5-15 kV/cm) ❖ Few relatively long duration pulses (>10 μs) ❖ Isotonic protective media

Figure 1.2 Pulsed electric field key components and variables.

5 μs range), and the medium was a salt solution or food. Reversible pulsed electric field effects relied on cell survival after PEF treatment for electroporation, and generally employed isotonic media and a few relatively low strength electric field pulses (~5–15 kV/cm) of relatively long duration (10 to hundreds of μs range).

Reversible electroporation was promoted by Zimmerman in Germany, who induced "pearl chains" of cells, that were then fused through PEF induced cell membrane effects (Zimmerman et al., 1976; Zimmerman, 1986). Zimmerman's contributions over many years played an early pivotal role, instrumental to the many cell biology and therapeutic developments that followed. PEF treatment for cell electroporation and genetic manipulation has since become an accepted and widespread method. The ongoing developments by Hofmann (personal communication) and others seem, in retrospect, a logical step in a direct line (followed by many workers) back to the contributions of Zimmerman. In the early 1980s, Hülshegher et al. (1983) published a series of papers on PEF microbial inactivation, stirring new interest and setting the stage for much of the work that followed.

PEF investigators studying inactivation and preservation effects have been highly inventive in treatment chamber design (Figure 1.3). The earliest chambers were designed to treat a confined, static volume. For example, some of the first designs incorporated parallel plate geometry using flat electrodes separated by an insulating spacer. One disadvantage apparent in this design is its inherent field strength limitation due to surface tracking on the fluid or insulator that leads to arcing. With interelectrode distances of ~1 cm, the onset of surface tracking and associated electrode pitting is observed at field strengths greater than ~22–24 kV/cm using 2.5–5 μs pulses.

This field strength limitation is design related; note that the liquid/insulator/electrode interface represents a triple-point, a fact that can be confirmed using breakdown test chambers employing radiused parallel plate electrodes with the insulator located in a low field region. Such chambers can operate

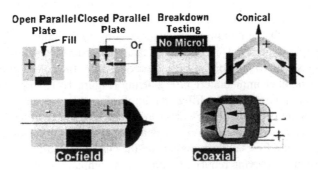

Figure 1.3 PEF treatment chambers.

at very high field strengths without experiencing breakdown; however, they are not generally useful for microbiological testing because of fluid exchange between regions receiving high and low levels of treatment. Current studies at the National Center for Food Technology using gelled media during PEF treatment in batch chambers may overcome this problem and add new insights. Parallel plate electrode designs naturally evolved to conical chamber shapes that offer the advantage of ease in eliminating bubbles, and are also easily modified to treat products' in-flow.

Two-chamber designs are currently favored for treating flowing product: coaxial (Bushnell et al., 1993; Qin et al., 2000) and cofield (Yin et al., 1997) arrangements. Coaxial chambers have inner and outer cylindrical electrodes with the product flowing between them. One or both electrodes are shaped so as to minimize field enhancements and grade the field into and out of the treatment zone. Coaxial chambers do not provide a uniform treatment across the treated volume since the radial geometry ensures that field strength will decrease towards the outer electrode; however, the degree of field nonuniformity can be predicted and controlled during chamber design. One challenge associated with coaxial chambers is that they generally present low load resistance when used to treat most foods, and the pulser system must be able to deliver high current at the voltages employed. Cofield chambers have two hollow cylindrical electrodes separated by an insulator so as to form a tube through which the product flows. Field distribution in a cofield chamber is also not expected to be uniform, though some useful advantages may be gained by special shaping of the insulator. The primary advantage of cofield chambers is that they can be designed to present high load resistance, and allow the pulser to operate at lower currents than those commonly employed with coaxial designs.

Nonuniformity in treatment weighs or favors the minimum response. This is inherent in the logarithmic nature of microbiological enumeration, and can mask the true dose response relationships of an organism to the process. For example, consider a situation in which treatment is 100% effective against 99% of the organisms in a sample but is degraded so as to show zero effectiveness against 1% of the organisms in the same sample. Maximum treatment effectiveness under these conditions is limited to production of a 2 log-cycle reduction, even if the treatment produces a 7 log-cycle reduction in the bulk of the sample. Therefore, in our batch chamber (parallel plate) studies, we withdrew a small volume of the treated product from the central region of the treatment chamber immediately after pulsing (Figure 1.4). This protocol was developed after we observed a different microbiological result when we emptied the entire chamber (with mixing) vs. the result obtained from collecting a small volume from the central region of the chamber. It was hypothesized that field grading or other effects at the periphery of the chamber resulted in a small volume of product receiving reduced treatment, as compared to the bulk chamber volume.

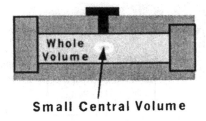

Small Central Volume

Figure 1.4 A PEF batch chamber sketch.

IN-FLOW PEF TREATMENT

When performing in-flow treatment studies it is important to avoid contamination of the lines downstream of the treatment chamber with untreated product. Such contamination is not easily washed out and will affect subsequent sample results. During startup this can be accomplished by pre-equilibrating the system with a sterile buffer matched in electrical properties with the product to be treated. The inlet line to the system is then switched from the sterile buffer to the test product without an interruption in pulsing.

For similar reasons, it is important to start dose response studies at the highest treatment level (Figure 1.5), i.e., the first data points collected should be produced by the highest treatment studied, and subsequent data points collected only after suitable equilibration time at the next lower treatment level. Frequently, it is appropriate to begin PEF treatment at a level resulting in chamber outlet temperatures known to produce thermal inactivation. In our studies we normally collected three sets of three samples at each treatment condition, usually over a period of 4–8 minutes, and then reduced the treatment level and allowed a similar period for equilibration before collection of the next set

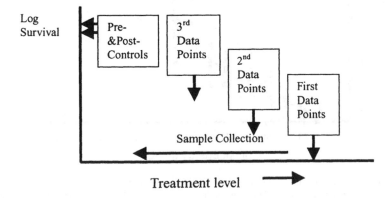

Figure 1.5 In-flow treatment test.

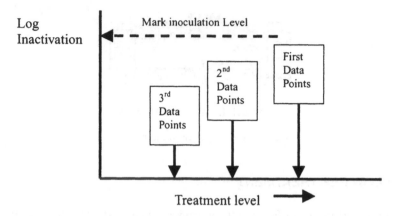

Figure 1.6 Microbial inactivation as a function of treatment level.

of samples. A similar number of pre-treatment samples (taken from the inlet product storage tank) and post-treatment samples (taken from the outlet of the system after suitable equilibration following the cessation of pulsing) were collected as controls. The results were then plotted as log survival [or its inverse expression (Figure 1.6), log inactivation] vs. the level of treatment employed.

Figure 1.7 shows the dose response obtained treating *Escherichia coli* inflow with a PEF treatment system fabricated at Pure-Pulse Technologies, San Diego, CA (Bushnell et al., 1993, 1996). This system can be operated with

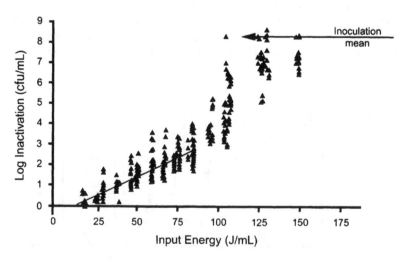

Figure 1.7 Inactivation kinetics of *E. coli* ATCC 26 varying e-field, 100 L/hr, 13 hz, ~12 pulses with inlet temperature of 40°C. The model nutritive medium used was: 0.5% tryptose, 0.3% beef extract and 0.3% dextrose. The conductivity was adjusted at 20°C to 250 Ω-cm with pH 7.0.

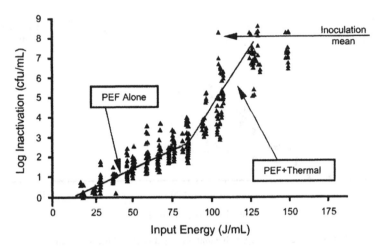

Figure 1.8 *E. coli* treated with PEF alone and PEF combined with heat.

one or two coaxial treatment chambers and has associated heating/cooling systems for control of the inlet temperature to each chamber, and for rapid cooling of treated product upon exit from either chamber. The results of trials performed on multiple occasions are shown plotted together (Figure 1.8). These results were obtained with only one treatment chamber. The test product was a model nutritive medium adjusted in resistivity, with pH 7.0 phosphate buffer to approximately match the conductivity of milk. The microbiological data is plotted as the log reduction obtained for a particular PEF treatment vs. the energy input (joules/mL) resulting from that particular PEF treatment. This energy input, based on the observed thermal excursion during processing, closely matched (~95% agreement) the energy input calculated from pulser capacitor charge voltage and stored energy. Thermal inactivation studies using this strain of *E. coli* do not show reduction in viability after 15 or 30 seconds holding at 60°C, however, similar holding times at 65°C show thermal inactivation. Our interpretation of the results suggests that the reduction observed below ~80 J/mL input energy (maximum treatment temperatures lower than 60°C) resulted from the effects of PEF treatment alone, and that the reduction observed at greater treatment levels and product temperatures (>80 J/mL input energy and >60°C product temperature) resulted from a combination of PEF and thermal effects. The inactivation curve shown in Figure 1.9 is, therefore, seen as comprised of two distinct regions of processing effects. Those values below 80 J/mL input energy represent the effects of PEF in the absence of normal thermal inactivation. We have chosen a linear regression model to represent the semilogarithmic plotting of the data (log reduction vs. input energy). This data is shown regressed separately from those values above 80 J/mL input energy and product temperatures >60°C.

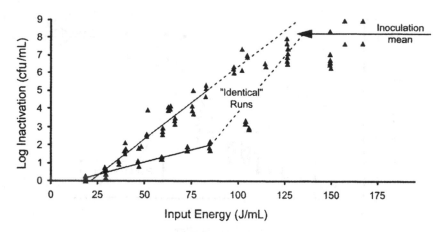

Figure 1.9 Influence changes in hydrodynamic fluid patterns on kinetics inactivation of *E. coli* ATCC 26.

The data shown in Figure 1.8 was obtained from treating *E. coli* on multiple occasions. A significant spread in the results between occasions is evident even though the inoculation level was held relatively constant. We believe this variation in the data is due to irregularities in flow, and hence level of treatment, through the chamber. We have observed, for example, that changes in piping configuration leading to the chamber inlet can result in significant changes in inactivation kinetics and in the level of microbiological effectiveness measured. We have also observed what we interpret as two extremes in *E. coli* inactivation kinetics (Figure 1.9). In some trials, little change in slope appears to occur in the transition from PEF to PEF + Thermal portions of the dose response curve. Whereas in other tests, a clear increase in inactivation per Joule of input energy occurs after about 80 J/mL of input energy. Insufficient data exists to decide which of these two extremes best represents the inactivation kinetics resulting from PEF treatment. The latter response is most frequently observed. However, if, as suggested, these changes in apparent inactivation kinetics arise from variations in fluid flow, it will be difficult to draw meaningful conclusions about the effectiveness and utility of PEF processing until these flow-dynamics issues have been understood.

PEF CAN BE APPLIED UNDER CONDITIONS THAT CONTROL POTENTIAL ELECTROCHEMISTRY

We can conclude that it is possible to control, reduce, or eliminate the production of chemical change in the processed product by tailoring the choice of PEF treatment conditions. Furthermore, under these reduced or null chemistry PEF processing conditions, we can demonstrate PEF killing effects using time and

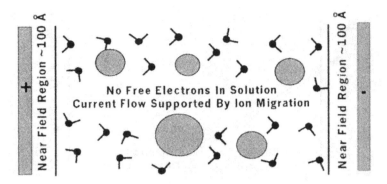

Figure 1.10 Ion migration in a parallel plate chamber.

temperature processing conditions well below those conditions where thermal inactivation would be expected to occur.

One core understanding of modern electrochemical science is that all electrochemical change occurs at or very near the surface of the electrodes, i.e., electrochemically induced electron exchange (reduction/oxidation) reactions occur only in an extremely small region, the "near field" region, at or within as little as 100 Å of the electrode surface (Figure 1.10). Thus, no free electrons are envisioned as flowing through the bulk fluid medium; current conduction through the medium is maintained via ion migration, and the circuit is completed by electron exchange reactions at the electrode.

It is also known that when an electrical pulse is applied to aqueous media, much of the energy early in the pulse (on a μs and sub-μs time scale) is consumed by orientation of water molecules (Figure 1.11). This phenomenon is associated with the double layer capacitive effect as energy is stored by forming oriented layers of water dipoles at each electrode. One effect of this phenomenon is a delay in the appearance of the full potential of the electrical pulse in the near

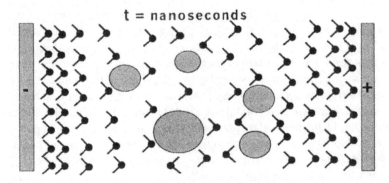

Figure 1.11 Orientation and distribution of ions and electrodes due to capacitive effect.

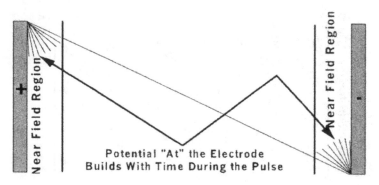

Figure 1.12 Electric-field in near field of the electrodes.

field region at the electrodes. Field potential at the solution/electrode interface is delayed, relative to the appearance of field strength across the cell, early in the pulse and increases with time during the pulse (Figure 1.12).

This delay in field potential in the near field region at the electrodes (the same region where all electrochemical reactions occur) means that insufficient electromotive force is available to initiate electron exchange reactions for some time during the pulse. This delay in potential can be quite dramatic with times of greater than about 1 μs required to reach a potential greater than one volt in the electrode near field region. Thus, it is possible, by selecting appropriate pulse duration, to apply PEF processing under conditions that prevent the occurrence of specific reactions for which the formal potential requirements are known; i.e., selection of pulse conditions for PEF processing allows the potential for electrochemical change to be minimized. For example, when applying a 30 kV/cm pulse to milk or products of similar conductivity, it is possible to maintain the potential at the electrode at less than the formal potential of the chloride ion by limiting pulse duration to less than about 2–2.5 μs, depending upon pulse shape (for example, rectangular vs. RC decay, rectangular pulses requiring shorter duration). However, during repetitive pulsing it is necessary to provide some mechanism to reset the electrodes to zero potential between pulses. We chose to perform our PEF processing trials in this "null electrochemistry" regime, and the experiments presented in this paper were performed using treatment conditions calculated to control and maintain electrode field potential at less than about the formal potential of the chloride ion. It should be noted that these conditions do not eliminate the potential for all electrochemical reactions, as some reactions will have formal potentials less than this value. However, they do limit the electrochemical reaction potential to those having formal potentials less than about that of the selected cut-off. The electrodes should, of course, be fabricated using materials demonstrating minimal reactivity under the processing conditions (i.e., "spectator" electrodes). In the experiments reported here, the electrodes were fabricated using graphite.

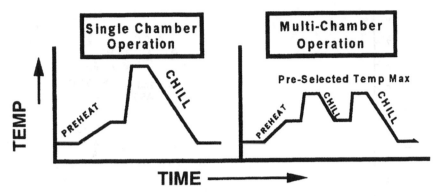

Figure 1.13 Multi-pass PEF processing vs. single pass.

MULTI-PASS PEF PROCESSING

Just as it is possible to control the potential for electrochemical reactivity during PEF processing, it is also possible to control temperature conditions throughout the process. This is easily accomplished by using multiple treatment chambers (Figure 1.13). When performing PEF processing with a single treatment chamber, the normal thermal cycle would include: (1) use of a heat exchanger to preheat the product before entering into the PEF treatment chamber to benefit from thermal synergy (there is a clear benefit to PEF processing at temperatures above refrigeration, between 40–45°C); (2) an increase in temperature as the product is PEF treated, and (3) use of a post-treatment heat exchanger to cool the product after treatment. There is a direct relationship between the electric energy supplied and temperature rise in the product. In this single treatment chamber mode the amount of PEF energy that can be applied under nonthermal conditions is limited, and the resulting microbial inactivation effects are difficult to distinguish in terms of time/temperature relationships from normal thermal inactivation.

To show the ability of the PEF process to produce inactivation effects significantly greater than would be expected based on time/temperature considerations alone, one must couple two or more treatment chambers in series (with cooling systems between chambers), so more PEF treatment can be applied under conditions allowing the maximum thermal excursion to be controlled below pre-selected values. The results of a two-chamber treatment can be simulated (Figure 1.14) using a single chamber in a multiple pass mode. This is accomplished by collecting a volume after a first pass through the system (at some particular maximum thermal excursion, 60.1°C in the example shown) and re-passing this sample through the PEF system (after CIP). In the example shown, the first PEF treatment pass (35 L/hr in-flow treatment at ~25 kV/cm) resulted in a thermal excursion from 20°C to 60.1°C, and achieved less than 1 log-cycle

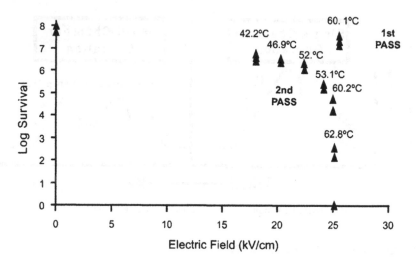

Figure 1.14 Single vs. double pass treatment.

reduction in *E. coli* survival. However, a second pass through the PEF system resulting in a thermal excursion from 20°C to 60.2°C, yielded about a 3 log-cycle reduction.

It was this type of finding which led us to fabricate a two-chamber PEF system. The results using a two-chamber system (Figure 1.15) confirmed earlier observations made using a one-chamber simulation. In the example shown, *E. coli* was treated in a two-chamber system with inlet temperature to each

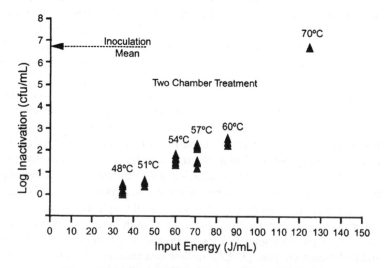

Figure 1.15 In-flow inactivation kinetics of *E. coli* ATCC 26.

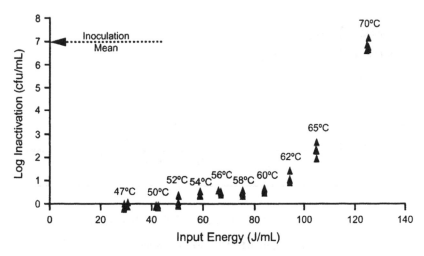

Figure 1.16 In-flow inactivation kinetics of *Listeria innocua.*

chamber controlled by heat exchangers to 40°C and the PEF treatment in each chamber controlled to the maximum temperatures as shown.

This beneficial use of multipass treatment is also seen in *Listeria innocua* PEF treatment studies (Figure 1.16). PEF treatment using a single chamber shows very little PEF inactivation until PEF processing yields temperatures where normal thermal inactivation is anticipated. *Listeria innocua* was chosen for study as a safer-to-handle surrogate for *Listeria monocytogenes*. The *Listeria innocua* tested had a thermal resistance profile similar to that of *Listeria monocytogenes.*

Time/temperature studies showed no thermal inactivation at 65°C with holding times of 15 or 30 seconds, but significant inactivation after exposure to 70°C for the same 15 or 30 second holding periods. *Listeria monocytogenes* is reported in the literature as having a 15 second holding decimal reduction value of ~69°C.

However, using multipass treatment through a two-chamber system to simulate treatment in 2, 4, and 6 chamber systems (through the use of the previously cited multipass experimental design), shows that 6 log-cycle reductions in *Listeria* viability can be achieved using PEF processing at a maximum temperature of 55°C. In the experiment shown in Figure 1.17, *Listeria innocua* was PEF processed in fresh unpasteurized, unhomogenized milk. The initial two-chamber PEF treatment (with maximum temperature of 55°C in each chamber) yielded just slightly over 1 log-cycle reduction, whereas a second pass yielded an additional reduction of ~3 log-cycles. A third pass through the two-chamber system reduced the recovery of the inoculated *Listeria,* and all other vegetative organisms present, to below the limits of detection—just slightly less than 100 colony forming units (cfu) per mL of heat resistant organisms (surviving heat shock at

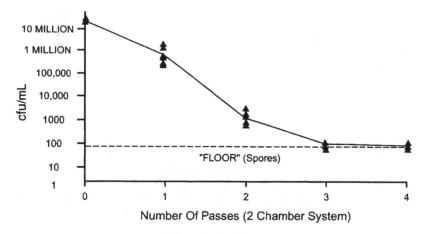

Figure 1.17 Multi-pass treatment.

85°C for 5 minutes and displaying bacillus-like colonial morphology). These were presumed to be spores initially present in the milk, and only these heat resistant organisms were recovered from the 6-chamber multipass treatments.

Other studies using *Listeria innocua* inoculated into milk and a model nutritive medium confirm that:

(1) PEF processing can produce significant levels of inactivation under treatment time and temperature conditions that would not be expected to yield inactivation through normal thermal mechanisms (Figure 1.18).

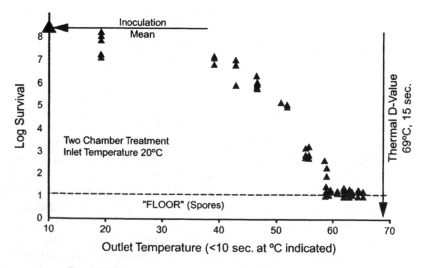

Figure 1.18 *Listeria innocua* inactivated in milk at low temperature.

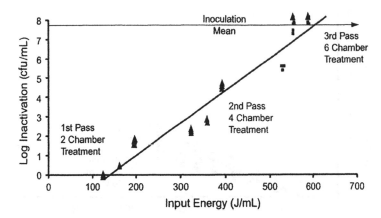

Figure 1.19 Multi-pass *Listeria innocua* ATCC 33090 inactivation kinetics. Nutritive solution (250 Ω-cm at 20°C, 100 L/hr, 13 Hz/cell, ~24 pulses/pass, 2 chamber treatment/pass with inlet temperature of 40°C.

(2) Multipass treatment can provide reduction levels greater than about 6 log-cycle in *Listeria* viability under these same nonthermal conditions (Figure 1.19). However, it is also shown that these successes are achieved only through the expenditure of considerable energy.

PEF PROCESSING USING NULL-ELECTROCHEMICAL CONDITIONS: FURTHER OBSERVATIONS

Next let's examine some features of PEF processing demonstrated during in-flow trials using treatment conditions calculated to minimize the potential for electrochemical change in the treated product. Using *E. coli* as a model gram-negative bacterium, studies in model nutritive medium adjusted to three different resistivities (250, 500, and 750 Ω-cm) suggest that, within this range of conductivity, product resistance does not significantly affect the level of microbiological kill obtained after PEF treatment (Figure 1.20). The regression lines shown in the chart include only those data points resulting in maximum product temperatures not greater than 60°C and, therefore, are taken to represent the effects of PEF treatment in the absence of added thermal kill. As was observed previously, the semilogarithmic depiction of the results obtained in each set of trials (in terms of the log kill measured vs. the input electrical energy producing the observed effect) appears to be best represented by a straight line. Our interpretation of the dose response displayed is that microbiological kill during PEF processing follows first order reaction kinetics in relation to energy consumption. Secondly, PEF processing appears to be able to produce this killing effect at time/temperature conditions that would not be expected to result in kill under normal heat treatment conditions. Finally, these kill kinetics do not appear to be

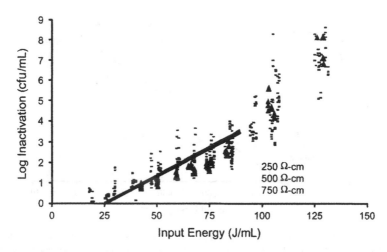

Figure 1.20 Influence of resistivity within range of 250 to 750 Ω over inactivation of *E. coli* ATCC 26.

affected by product resistivity within the range tested, 250–750 Ω-cm. There are reports in the literature that differ from this conclusion. It should be pointed out that the number of pulses applied to reach a certain energy input level does vary significantly during PEF processing at each resistivity. Higher resistivity products require more pulses to reach a particular input energy value than products of lower resistivity treated under the same conditions, since lower energy per pulse is delivered at higher resistivity (all other treatment parameters held constant).

Studies were conducted to investigate the potential of the PEF process to inactivate organisms in particulates (Figure 1.21). Model particles containing *E. coli* were formed by mixing the organisms into a Na-alginate solution and then droppering this mixture into $CaCl_2$ to form alginate beads containing *E. coli*. These beads were readily solubilized by mixing in sodium pyrophosphate, and control experiments showed the bead formation and solubilization process did not effect *E. coli* viability. Figure 1.21 summarizes the results obtained when beads designed to be of relatively low or high resistivity were PEF processed, solubilized, and *E. coli* survival evaluated. The lethal effects of PEF treatment on organisms extracted from the beads are shown compared with our database for PEF effects against *E. coli* treated in medium. We interpret the results obtained as indicating that very little difference exists between PEF effects on *E. coli* suspended in medium and *E. coli* entrapped in the model particulates. We believe the small differences recorded in killing in the beads, vs. that in medium, relate to slightly higher PEF exposure in the beads. This is because their residence time in the treatment chamber was slightly extended by flow artifacts associated with bead movement through the chamber (drag along the chamber walls during transit).

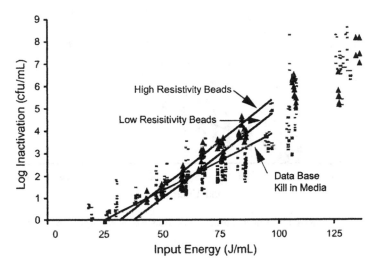

Figure 1.21 Inactivation in model particles of *E. coli* ATCC 26 in 2 mm diameter alginate beads in model nutritive media.

Our earliest tests treating organisms in static, batch PEF chambers indicated a benefit could be obtained by preheating the treated product, even though the preheating process and temperatures did not produce lethal effects, and though the final PEF treatment conditions were controlled so the maximum temperature and holding time would not be expected to yield thermal killing effects. This thermal synergy has since been observed in other laboratories. Figure 1.22 demonstrates this effect against the indigenous microorganism population in juice observed during in-flow PEF treatment of fresh squeezed orange juice obtained from a commercial source. A benefit to preheating the juice to 40°C is

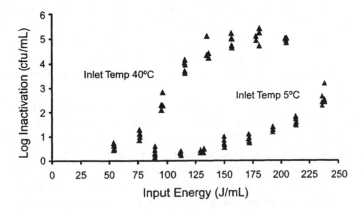

Figure 1.22 Thermal synergy in fresh squeezed orange juice.

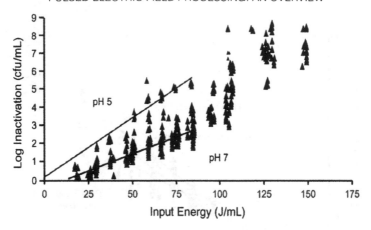

Figure 1.23 Effect of pH over inactivation of *E. coli* ATCC 26.

apparent in the results obtained. This benefit was seen even in those samples in which normal thermal inactivation (maximum holding time <15 seconds) is not expected. (Though the PEF inactivation effects are plotted against input energy rather than the resultant temperature, one can calculate the approximate temperature reached during treatment for each sample by assuming a heat capacity of ~4 J/mL per 1°C change in temperature.) In the example shown, a mixed population of organisms was treated; this affects the kinetics (log reduction vs. energy input relationship) observed. A PEF inactivation pattern often seen in treating homogeneous populations of a model organism at different inlet temperatures is parallel curves between the two different test temperatures, the lower inlet temperature sample curve being displaced to the right (higher energy input) but symmetrical with the higher inlet temperature results (data not shown).

The pH of the medium can have an effect on the PEF killing effects obtained. Figure 1.23 compares the inactivation kinetics obtained treating *E. coli* in-flow in a model nutritive medium adjusted to pH 5 and pH 7. PEF processing of *E. coli* is more efficient at a pH of 5 than at a pH of 7. Control experiments show the *E. coli* strain used in these tests does not decrease in viability when exposed to the pH 5 medium under the time and temperature conditions employed in the absence of PEF treatment.

The same *E. coli* strain did not show a difference in inactivation kinetics when the results of PEF treatments in media adjusted to pH 6 and pH 7 were compared (Figure 1.24). *Listeria innocua* was also tested as a model gram-positive microorganism in media adjusted to various pH values. In Figure 1.25, PEF inactivation effects obtained treating *Listeria innocua* in media at pH 5 and pH 7 are compared. With regard to pH effects it should be noted that although the two examples shown appear to respond similarly to changes in pH, we have observed that other organisms may display different response patterns.

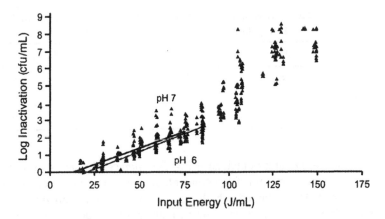

Figure 1.24 Effect of pH over inactivation of *E. coli* ATCC 26.

PEF INACTIVATION OF HEAT RESISTANT MOLDS (ASCOSPORES)

The survival of heat resistant molds, particularly ascospores which are in general more resistant than vegetative forms, in thermally processed foods and juices has long been a food processing challenge and problem. In order to investigate the potential effects of PEF processing on heat resistant molds, ascospores of the fungus *Talaromyces flavus* were inoculated into pineapple juice. *Talaromyces flavus* ascospores are quite resistant to thermal treatment. The thermal death rates and the calculated decimal reduction times (*D* value, time of exposure at a temperature yielding one log-cycle reduction in survival) obtained during treatment at 80°C, 85°C, and 90°C are shown in the

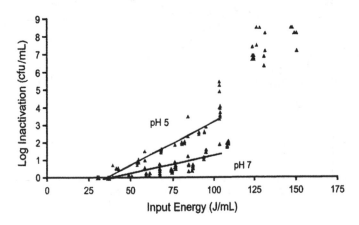

Figure 1.25 Effect of pH over inactivation of *Listeria innocua* ATCC 33090.

TABLE 1.1.   Heat Resistance of *Talaromyces flavus* Ascorpores: *D*-Value Times for 1 Log Reduction in SS-Pineapple Juice.

	pH	90.6°C (195°F)	98.9°C (210°F)
Trial 1	3.8	23 min	0.6 min
	4.2	7.9 min	0.9 min
Trial 2	3.8	12 min	0.9 min
	4.2	6.7 min	0.6 min

Figure 1.26. The *D* values obtained using *Talaromyces flavus* ascospores inoculated into single strength (from concentrate) pineapple juice are shown in Table 1.1. The values obtained in both figures correlate well with values in the literature; *Talaromyces flavus* ascospores require long exposures to temperatures near boiling to produce significant reductions in viability.

To evaluate the effects of PEF processing against this heat resistant mold, *Talaromyces flavus* ascospores were treated in pineapple juice using a multipass PEF treatment regime. The results obtained after one and two passes through a two-chamber system (four treatment cycles after the second pass) are shown in the Figures 1.27 and 1.28. Though ~3.5 log cfu of ascospores were originally inoculated, ~4.5 log cfu were recovered from the least treated samples collected after passage through the PEF system. This apparent increase in cfu is not unusual when working with ascospore populations and is explained by disaggregation of clumped cfu (this effect is most often due to damage to the ascus membrane and the release of spores previously bound within asci). The residence time of the product within each PEF chamber until it was cooled by

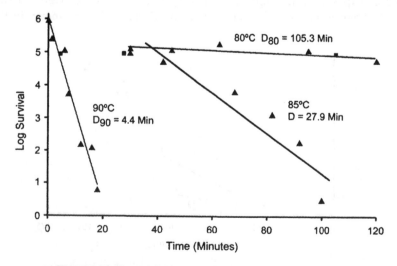

Figure 1.26 Thermal death rate of *Talaromyces flavus* ascorpores.

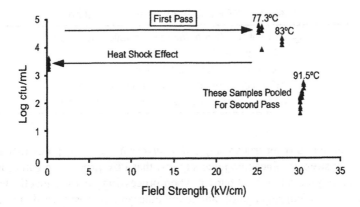

Figure 1.27 *Talaromyces flavus* ascorpores in SS-pineapple juice.

the downstream heat exchanger was determined to be less than 5 seconds, so the residence time at the maximum temperature resulting from PEF processing during one pass through the system was less than 10 seconds. During the first passage through the system, a large volume of product was collected during PEF treatment that resulted in a maximum temperature of 91.5°C. This sample was then re-passed through the system (after CIP). The inactivation kinetics seen during the first passage through the PEF system clearly indicate significant effects in excess of those expected based upon the time/temperature conditions of exposure; i.e., the inactivation seen is greater than that expected through normal thermal effects alone. This conclusion is reinforced and extended by the results observed during the second passage through the PEF system.

The conclusions regarding the inactivation levels seen during PEF processing vs. strictly heat-associated effects are summarized in Table 1.2. The inactivation

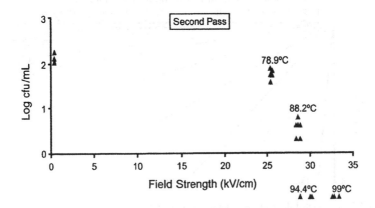

Figure 1.28 *Talaromyces flavus* ascorpores in SS-pineapple juice.

TABLE 1.2. **PEF and Ascospore Inactivation.**

❖ Thermal Time/Temperature Inactivation Conditions:
  • 3.5 logs ⟶ 97°C for 4.1 min
  • 3.5 logs ⟶ 92°C for 27.4 min
❖ PEF Time/Temperature Inactivation Conditions:
  • 3.5–4.5 logs ⟶ 85–95°C for less than 0.3 min

seen in these and other trials using *Talaromyces flavus* ascospores inoculated into single strength pineapple juice suggest that PEF processing can dramatically reduce product thermal burden when processing juices to inactivate heat resistant molds. Pineapple juice (and many other juices) possesses delicate sensory attributes that are often thermally sensitive and degrade to off-notes. It is quite possible that the increased inactivation levels obtained using PEF processing vs. thermal treatment, can provide a desirable sensory benefit. However, it must be pointed out that a high level of energy expenditure may be required to take full advantage of this benefit.

PEF INACTIVATION OF BACTERIAL SPORES

In experiments treating bacillus spores under low temperature conditions (<100°C), no inactivation was observed. In order to study the effects of combined thermal and PEF processing on bacillus spores, a *Bacillus stearothermophilus* spore population was obtained from the National Food Processors Association/National Food Laboratory in Dublin, CA. This *Bacillus stearothermophilus* spore preparation was characterized as to thermal response by the

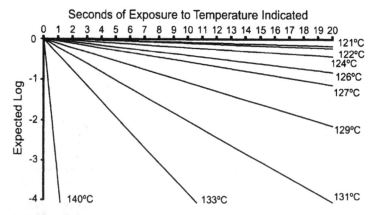

Figure 1.29 Calculated thermal inactivation kinetics of *B. stearothermophilus* NFPA/NFL 7–85 spores with $D_{121°C} = 120$ s and $Z = 7.2°C$.

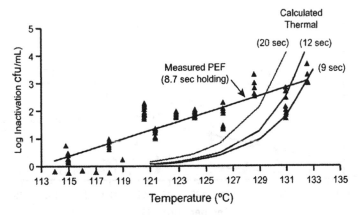

Figure 1.30 PEF inactivation kinetics of *B. stearothermophilus* NFPA/NFL 7–85 spores treated in 250 Ω-cm at 20°C and pH 7.

NFPA/NFL; it had a $D_{121°C}$ of 120 seconds and a Z of 7.2°C. Figure 1.29 shows a family of time/temperature inactivation curves calculated for thermal inactivation of an organism with these D- and Z-values.

The inactivation observed during passage through a two-chamber PEF system is shown in Figure 1.30, along with the kill kinetics expected based upon strictly thermal inactivation for holding times of 9, 12, and 20 seconds. The experiment was performed in a phosphate buffer (to minimize the potential for spore germination induction by organic stimuli) adjusted to an electrical resisitivity similar to that of milk. In-flow treatment was performed at 75 L/hr using an inlet temperature of 70°C to each PEF treatment chamber. The PEF associated thermal cycle holding time was calculated to be 8.7 seconds. This was the combined time from entry into each of the two PEF chambers until a 5°C cooling was accomplished in the heat exchanger downstream from each PEF chamber. This time represents the calculated holding time at the maximum temperature reached during PEF processing (the come-up time to this maximum temperature is fractions of a second in each PEF treatment chamber). A significant difference in inactivation kinetics is observed between the measured PEF response and the calculated thermal inactivation curves. Though the fit of the PEF data to the linear regression shown is relatively imprecise (R^2 of 0.87), the trend in the PEF measurements does seem best fit to a linear model in the semilogarithmic depiction of kill vs. final temperature during treatment (the final temperature is proportional to energy input). Whereas the calculated thermal kinetics curves, at the temperatures involved, show only the toe of what will become a relatively steep thermal death response. The differences in temperature, slope, and shape between the observed PEF inactivation kinetics and those expected from normal thermal inactivation suggest that significant differences exist between the two processes.

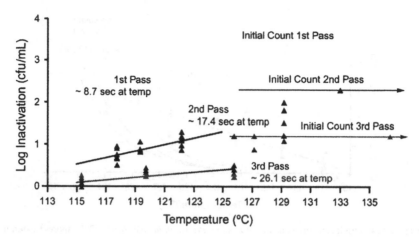

Figure 1.31 Multi-pass inactivation vs. outlet temperature of *B. stearothermophilus* NFPA/NFL 7–85 spores.

Figures 1.31 and 1.32 show the results obtained after PEF treatment of the *Bacillus sterarothermophilus* spores using multipass treatment through a two-chamber system. The results are shown plotted as the log inactivation obtained vs. the maximum treatment temperature (Figure 1.31) or the PEF processing energy input (Figure 1.32).

As we have already seen during the PEF treatment of heat resistant mold spores, the results obtained using *Bacillus stearothermophilus* spores show that PEF processing can be used for inactivation under reduced time/temperature treatment conditions relative to those required during normal thermal treatment.

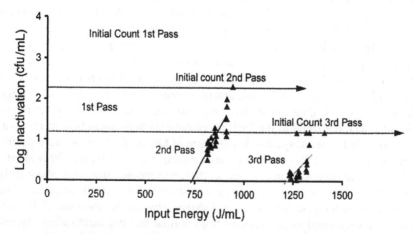

Figure 1.32 Inactivation of *B. stearothermophilus* NFPA/NFL 7–85 spores.

Other interesting observations and comments can be made relative to the results demonstrated, however, one important point seems apparent: the energy expenditures associated with the spore inactivation results achieved in these tests appear daunting.

PEF EFFECTS ON ENZYMES AND VIRUSES

The literature is mixed and confusing with regard to the potential effectiveness of PEF treatment on enzymes and viruses in treated products and model solutions. Using several model enzyme systems, and bacteriophage as a viral model, we performed tests in static batch treatment chambers and in products treated in-flow. In no instance have we observed PEF inactivation of either enzymes or viruses under conditions suggesting that mechanisms other than normal thermal inactivation are present.

FINAL REMARKS AND HYPOTHESES

Though we started with a selective historical overview of PEF development in general, the major focus of the discussion has been the microbiological killing effects of PEF processing. The results and findings observed in our laboratories at Pure-Pulse Technologies during programs supported by the United States Army and administrated through the RD&E Center at Natick Laboratories, Natick, MA, have been presented as specific examples.

There is much to be discussed in regard to equipment and engineering issues. It is paramount to note that the systems and equipment used by the individual laboratories reporting on PEF studies are often quite dissimilar, and these differences are in most instances both important and fundamental. They vary over a broad range including, but by no means limited to, differences in chamber design features and pulser and pulse characteristics. It is important to keep these differences in equipment and conditions in mind when comparing PEF results from individual laboratories throughout the world; often they are so wide ranging as to make direct comparisons difficult.

Although there are many differences in equipment and results, some important similarities in conclusions are evident in published reports and personal communications from numerous laboratories. PEF can clearly provide enhanced inactivation at lower time/temperature conditions than thermal processes. However, several opportunities for improvement remain, such as: (a) process energy consumption must be reduced, and (b) flow hydrodynamics and treatment uniformity must be understood and improved. These conclusions are also supported by the results reviewed in this report. Thus, it seems clear that PEF processing has proven it can live up to the promise predicted by early workers in the field, that of being able to inactivate organisms using time/temperature conditions significantly less than those employed during normal heat treatment. However, it seems equally apparent that accomplishment of these PEF reduced temperature killing effects are currently accompanied by significant energy expenditure requirements.

Commercial thermal processes (for example, milk pasteurization) are normally carried out using heat regeneration conditions (cooling the finished product by exchanging its thermal load to preheat the inlet product steam) that accomplish high overall thermal processing efficiencies (often greater than ~75%). Though some aspects of this approach can be utilized during commercial PEF processing, the bulk of the electrical energy input is not recoverable. Using the data presented here as an example, the electrical energy requirements for effective PEF processing are very high. Unless improvements in PEF treatment efficiency are forthcoming, it is doubtful the method will be commercially successful for the processing of foods.

However, outstanding questions remain concerning the efficiency of PEF application with current system designs, and understandings arising as we address these challenges may well lead to solutions impacting this issue. At several points during the discussion herein we have tried to emphasize the paramount importance of treatment uniformity and the profound impact nonuniform treatment (arising from either field variation throughout the treated volume or irregularities in flow during processing) can have on kill kinetics and understanding the PEF process.

The dose response of microorganisms to PEF treatment is often seen as composed of two approximately log-linear segments with an increase in slope at treatment levels, resulting in time and temperature conditions where organisms are known to transition to thermal death. In tests using *E. coli* and in-flow PEF treatment, this transition occurs at treatment levels yielding maximum temperatures greater than about 60°C; for *Listeria innocua* the transition occurs at just above 65°C. It is noteworthy that at time/temperature conditions less than this thermal death transition point, PEF treatment is able to produce microbiological kill, and a log-linear relationship appears to exist between the microbiological effects produced by PEF treatment and the energy input employed. A high degree of variability in the slope of both segments of the kill curve is seen, however, from experiment to experiment, even though organism preparation and inoculation are held relatively constant. The results of presumably identical PEF treatments can vary from about one log-cycle reduction to >4 log-cycle reductions for the PEF only portion of the inactivation response, and may show even greater imprecision for the PEF + thermal inactivation seen during PEF treatments above the thermal transition point. High variability at temperatures above the thermal transition point is not seen using the PEF system and its associated heat exchangers to generate thermal death curves with the electrical pulser turned off. This high variability seems to be due to nonuniformities in flow or exposure during PEF treatment.

Two kinds of nonuniformity of flow and treatment exposure were noted in the PEF trials presented here: (1) variation in treatment level within the collection of assayed samples due to fluctuations in bulk flow rate, and (2) geometric variation in flow rate or the level of treatment within a sample volume in the chamber. The former type of variation primarily results from flow rate instabilities during the collection of multiple samples. The in-flow PEF treatment system used in this study incorporated a Tri-Clover lobe pump, pulsation-dampener, and backpressure control valve (the system was normally operated under ~20–30 psi backpressure). A magnetohydrodynamic flow monitor was interfaced with the pump speed controller through PID feedback control. Each collected sample within the three sets of three samples taken at a particular treatment level was 50 mL in volume (equivalent to about two times the volume of the PEF chamber). Pump rate did vary slightly during the 4–8 minutes the 9 × 50 mL samples were collected. This variation in pumping rate was observable as a small variation in maximum treatment temperature during the time the multiple samples were collected. The normal range of variation in maximum treatment temperature observed during the collection of samples was less than about 1.5°C (equivalent to less than about 6 J/mL variation in PEF energy deposition per chamber). This flow variation also adds to the amount of scatter seen within the data points at a particular treatment level. However, this type of variation in treatment within a sample, due to pump or bulk flow rate fluctuation, is seen as being relatively small in magnitude and controllable by system design and operation.

A second type of potential nonuniformity in flow or exposure, geometric variation within the chamber volume, can be described as variation in terms of position within the roughly cylindrical coaxial treatment volume: azimuthal (around the cylinder), longitudinal (along the cylinder), or radial (across the gap) variation in treatment. It is this type of variation that may potentially result in greater variation across a sample volume than the small (~6 J/mL) variation recorded during bulk flow rate fluctuation. We have observed that the treatment chamber piping arrangement can significantly affect the inactivation kinetics obtained. It was hypothesized that some piping configurations led to the product "jetting" through the treatment chamber such that different flow rates (channeling) occurred around (azimuthal to the longitudinal axis of the electrode) the chamber. We also observed that chamber orientation (horizontal vs. vertical) can affect the results obtained, perhaps through the influence of convection related events. In some respects a PEF chamber can be considered analogous to a heat exchanger; significant energy is exchanged into the product in the relatively short distance and time during transit through a PEF treatment chamber. In our coaxial PEF chamber, a product is often electrically heated over seconds, or fractions thereof, many tens of degrees centigrade during a few centimeters of travel along the electrode. However, electrical energy deposition is not expected to be uniform across or around the treated volume. The radial design of the two electrodes dictates that the field strength, and hence energy deposition and heating rate, will be some percentage less at the outer margin of the chamber volume as compared to that at the inner margin. This differential heating rate will produce a change in conductivity across the product volume, as product resistivity normally falls with increase in temperature. Consequently, it is anticipated that this differential energy deposition (and heating) would be exaggerated with respect to subsequent PEF treatment pulses. A similar phenomenon is also anticipated along the electrode longitudinal axis, as the product becomes warmer and hence more conductive during travel along the electrode. The effects of such energy grading across and along the electrode are unknown. One of the electrodes is also tapered (or radiused) at each end to grade the field into and out of the treatment volume and prevent localized field enhancements. This may, however, produce eddying flow patterns or other flow irregularities that necessitate a trade-off between the optimization of flow and electrical performance properties of the chamber. Flow eddy patterns would tend to recirculate or trap product in the graded-field region leading to potential overtreatment. Relatively slow flow rates and laminar flow conditions were employed to obtain the experimental results presented here. Under these conditions, and with the problems and irregularities noted, it is not difficult to envision a variety of nonuniformities in flow and exposure effects, accompanied by convective or other events, as able to affect the kinetics observed. These potential irregularities, though believed to be present but not well understood, may contribute to pessimistic energy efficiency predictions.

While recognizing the importance of these outstanding questions and concerns, and the degree to which new understandings and process designs may provide new findings and insights, the data currently available has lead us to some hypothetical proposals. These hypotheses relate to the potential effects of a cell on field distribution during an electrical pulse. The cell, because of its dielectric properties in relation to those displayed by the bulk medium, is envisioned as producing a perturbation in electrical field and current distribution during the pulse. Thus, during the high intensity electrical pulse, localized field enhancements and current pinching may occur on the cell surface. Such localized field enhancements in the regions of a cell have been modeled at Washington State University (Spencer et al., 1996) and are potentially large. If the cell, because of its size and dielectric properties, is capable of producing significant changes in field and current distribution patterns during the pulse, it is not unreasonable to expect associated thermal and/or electrochemical events. From an electrochemical perspective, the cell surface might be envisioned as additional spurious electrode surface, as localized field effects lead to electromotive differential about the cell. Recently, it has been observed that product formation during electrochemical processing can be increased by incorporating microbeads into the reaction mixture, and it can be hypothesized that cells may play a similar role during PEF treatment. From a thermal perspective, it is not unreasonable to anticipate field enhancements and current pinching might lead to portions of the cell surface experiencing greater heating or thermal effects than the bulk medium. Such a selective heating phenomenon could produce thermal effects on the cell at time/temperature conditions in the bulk medium less than those predicted as necessary for normal thermal kill.

In summary, pulsed electric field methods have experienced considerable success and hold further promise in a variety of applications areas. PEF methods have gained wide acceptance for electroporation and cell biology applications. Recent results providing additional documentation of the potential benefits of combined pulsed electric field and chemotherapeutic treatments, hold further promise for PEF methods in medical science. This chapter has focused on our experiences in developing PEF methods for microbial disinfection and preservation effects. For a variety of organisms and applications, the method has been shown to be able to deliver on the promise, predicted by early workers in the field, of being able to provide microbial inactivation using processing time/temperature conditions significantly less than those required for similar effects during normal heat treatment. However, these successes required the expenditure of significant energy with the processing conditions and equipment employed. In this application, unlike the situation for many other PEF applications, processing energy efficiency is expected to be important for full acceptance. The development of equipment and understandings allowing greater effectiveness per Joule of energy input is the challenge that lies ahead.

REFERENCES

Bushnell, A. H., Dunn, J. E., and Clark, R. W. 1993. High pulsed voltage systems for extending the shelf life of pumpable food products. U.S. Patent 5,235,905.

Bushnell, A. H., Clark, R. W., Dunn, J. E., and Lloyd, S. W. 1996. Process for reducing levels of microorganisms in pumpable food products using a high pulsed voltage system. U.S. Patent 5,514,391.

Dovenspeck, H. 1960. Verfahren und vorrichtung zur gewinnung der einzelnen phasen nus dispersen systemen. German Patent 1,237,541.

Hamilton, W. A. and Sale, A. J. H. 1967. Effects of high electric fields on microorganisms II. Mechanism of action of the lethal effect. *Biochim. Biophys. Acta* 148, 789–800.

Hülsheger, H., Potel, J., and Niemann, E. G. 1983. Electric field effects on bacteria and yeast cells. *Radiat. Environ. Biophys.* 22, 149–162.

Qin, B-L., Barbosa-Cánovas, G. V., Swanson, B. G., Pedrow, P. D., Olsen, R., and Zhang, Q. 2000. Continuous flow electrical treatment of flowable food products. Part III. U.S. Patent 6,019,031.

Sale, A. J. H. and Hamilton, W. A. 1967. Effects of high electric fields on microorganisms. I. Killing of bacteria and yeast. *Biochim. Biophys. Acta* 148, 781–788.

Spencer, D. B., Pedrow, D. B., Olsen, R. G., Barbosa-Cánovas, G. V., and Swanson, B. G. 1996. Space change evolution in dilute binary electrolyte exposed to high voltage transients. *IEEE Transactions on Dielectrics and Electrical Insulation* 3(6), 747–753.

Sitzmann, W. 1990. Keimabtotung mit hilfe elecktrischer hochspannungsimpulse in pumpfahigen nahrungsmitteln. Vortrag Anlablich des Seminars "Mittelstansfourderung in der Biotechnologie." Ergebnisse des Indirekt-Spezifischen Programma des BMFT 1986–1989. *KFA Julich*, 6–7 Feb.

Sitzmann, W. 1995. High voltage pulse techniques for food preservation. In *New Methods of Food Preservation* (G. W. Gould, Ed.), p. 236. Blackie Academic & Professional, London.

Yin, Y., Zhang, Q. H., and Sastry S. H. 1997. High voltage pulsed electric field treatment chambers for the preservation of liquid food products. U.S. Patent 5,690,978.

Zimmermann, U. 1986. Electrical breakdown, electropermeabilization and electrofusion. *Rev. Phys. Biochem. Pharmacol.* 105, 176–256.

Zimmermann, U., Pilwat, G., Beckers, F., and Riemann, F. 1976. Effects of external electrical fields on cell membranes. *Bioelectrochem. Bioenergetics* 3, 58–83.

CHAPTER 2

Engineering Aspects of the Continuous Treatment of Fluid Foods by Pulsed Electric Fields

S. ESPLUGAS
R. PAGÁN
G. V. BARBOSA-CÁNOVAS
B. G. SWANSON

ABSTRACT

E NERGY and mass balances for the continuous single pass and recirculation mode of a pulsed electric field (PEF) installation were analyzed. The importance of the electric field and frequency pulses in food temperature increases was also discussed. Inoculation studies with *Bacillus subtilis* var. *niger,* ATCC 9372, in simulated milk ultrafiltrate (SMUF) were carried out in a continuous single pass operation to predict the results of recirculation. Good agreement between the experimental results with recirculation (total recirculation of the fluid leaving the PEF chamber) and the values predicted by the model were achieved.

INTRODUCTION

The increase in quality of foods constitutes one of the most important objectives in the food industry. Foods are normally treated using thermal processes, but in order to maintain their like-fresh quality, pasteurization is necessary by nonthermal processes (Barbosa-Cánovas et al., 1995). Among these, pulsed electric fields (PEF) may constitute one of the best preservation methods for fluid foods. The PEF treatment is conducted at an ambient or refrigerated temperature for a duration of microseconds, which preserves the fresh physical, chemical, and nutritional characteristics of foods. The inactivation of microorganisms is achieved by the effect of the electric field rather than by electrically induced thermal effects.

The application of electric fields to cellular suspensions produces a large increase in the conductivity and permeability of the membrane (Sale and Hamilton, 1967, 1968; Kinosita and Tsong, 1977; Zimmermann et al., 1980; Hülsherger et al., 1983). Depending on the potential of the electric field applied, breakdown or lysis of the membrane, with the consequent inactivation of the microorganism, can be achieved.

The application of PEF to a fluid may be a batch or continuous operation. In a batch operation, the food is first placed in the PEF chamber, and later several high voltage pulses are given to the food. However, the most interesting operation from a technological point of view is the continuous operation (Zhang et al., 1995), in which the fluid food flows continuously through a PEF chamber where the electric field pulse is applied.

EXPERIMENTAL INSTALLATION

The experimental installation consisted of a high voltage pulse generation unit, a feed reservoir that contains the fluid to be treated, one pump, a treatment chamber (PEF), and a treated food tank.

The high voltage pulse generation unit gives exponential decay pulses by discharging a capacitor of 0.5 μF into the PEF chamber, as shown in Figure 2.1. The power supply provided a maximum charging voltage of 40 kV. By using a specific pneumatic system commanded by one personal computer, exponential decay pulses (Figure 2.2) were generated and monitored with an oscilloscope (100 MHz digital HP54501).

The treatment chamber consisted of two electrodes (Figure 2.3) made of stainless steel. The inner electrode was cylindrical and connected to a high voltage supply and the outer electrode to the ground. The electrodes were separated with insulating material (polysulfone), that provided an annular chamber with an homogeneous electric field (Qin et al., 1995).

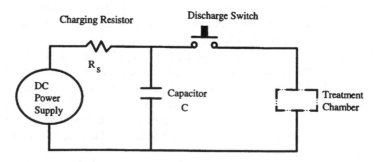

Figure 2.1 Simplified electric circuit for pulse generation.

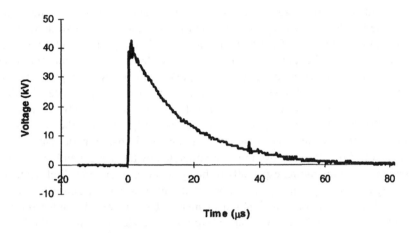

Figure 2.2 Voltage-time for an exponential decay pulse.

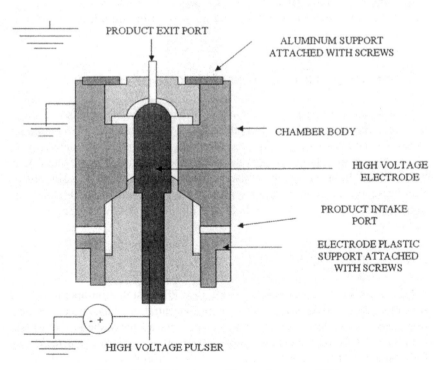

Figure 2.3 PEF chamber used in a continuous operation.

The outer electrode was connected to the ground, whereas the inner was connected to the high voltage source. The radius of the inner electrode was 20 mm, and the thickness of the PEF chamber was 6 mm giving a total volume of 26 cm^3. The PEF chamber was perfectly sealed to prevent the entrance of air bubbles, which would generate a spark followed by an explosion that could damage the PEF chamber. As with air bubbles, the presence of solid particles could also cause problems. The fluid food continuously enters and leaves the PEF chamber, which serves to aid the refrigeration of the electrodes and the reaction system. The chamber may operate in batch mode, but in this case it was difficult to remove the heat generated by the system and operate at room or low temperature, especially if the system was operated at high voltage and high frequency pulsing. Resistors in the electric circuit eliminated part of this heat generation. The amount of heat to be removed depended on the conductivity of the fluid food. Depending on the conductivity of the fluid and the operating conditions (voltage and frequency pulsing), it was sometimes necessary to refrigerate the inlet fluid food.

The experimental installation may be operated continuously with only one pass for the fluid food or with total recirculation of the fluid leaving the PEF chamber.

MATHEMATICAL MODEL OF CONTINUOUS OPERATION

From an engineering point of view, it is of interest to differentiate the single pass and recirculation modes. In both cases, the mathematical model consists of energy and mass balances, kinetic equations, and equilibrium conditions. It is possible to build a large and complicated mathematical model, but that would not be useful. In order to simplify the mathematical model, some assumptions may be adopted. Accordingly, plug flow in the PEF chamber, perfect mixing in the tank, and a first order kinetic for the elimination of microorganism was assumed.

SINGLE OPERATION

Figure 2.4 shows a simplified scheme of a PEF installation operating in a single pass mode. It can be assumed that the concentration of the microorganism in the feed tank c_T (microorganisms/L) is the same as that at the PEF chamber inlet, and that the rate of microorganism destruction r [microorganisms/(L/s)] follows a first order kinetics with respect to microorganism concentration c (microorganisms/L):

$$r = -kc \tag{1}$$

where $k(s^{-1})$ is the kinetic constant of microorganism elimination.

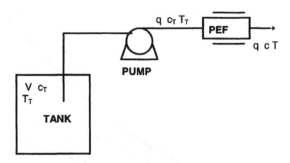

Figure 2.4 PEF in a single pass operation.

Assuming stationary state and plug flow in the PEF chamber (Levenspiel, 1972), the microorganism balance gives the following expression:

$$q \ln(c/c_T) = -kV_r t \tag{2}$$

where $q(L/s)$ is the fluid flow and $V_r(L)$ is the PEF chamber volume.

According to the last equation, the relation between the outlet microorganism concentration c (microorganisms/L) and time $t(s)$ is exponential.

$$c = c_T \exp(-kV_r/q) \tag{3}$$

The energy balances are more complex. The energy E (J) dissipated during the discharge of the capacitor $C(\mu F)$ at a voltage V (V) is given by the following equation:

$$E = 0.5CV^2 \tag{4}$$

Taking into account the frequency $f(s^{-1})$ of the pulses, the energy flow dissipated Q (J/s) is:

$$Q = f\,0.5CV^2 \tag{5}$$

However, only one part ϕ of this energy will heat the fluid food (flow q (L/s), density) that passes through the PEF chamber. This ratio ϕ must be less than 1, and depends strongly on the electrical conductivity of the food. The application of the energy balance to the PEF chamber when the stationary state is reached leads to:

$$q\rho C_p (T - T_T) = \phi Q \tag{6}$$

where q (L/s), ρ (kg/L), and C_p (J/kg°C) are the flow, density, and specific heat of the fluid food, respectively.

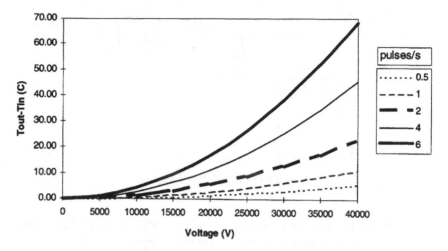

Figure 2.5 Increase of fluid temperature as function of operating voltage.

Consequently, the increase of the temperature $T - T_T$ of the fluid food will be:

$$(T - T_T) = \phi f 0.5 C V^2 / (q \rho C_p) \tag{7}$$

According to Equation (7) and Figure 2.5, it is possible to appreciate the strong influence of the applied voltage on the heat released to a fluid food. Figure 2.5 shows the maximum values achieved ($\phi = 1$) at several frequency pulses and voltages for a liquid with properties similar to water at a flow rate of 0.5 L/min [$C_p \neq 4200$ J/(kg°C), $\rho = 1000$ kg/m^3].

RECIRCULATION

Figure 2.6 shows a simplified scheme for processing foods in a recycling mode. Similar to the single pass operation, it can be assumed that the microorganism concentration in the feed tank $c - r$ (microorganisms/L) is the same as that in the PEF chamber inlet, and the rate of elimination of microorganism r [microorganisms/(L/s)] is first order with respect to its concentration c (microorganisms/L). The fundamental difference between the two operations in this case is that both the microorganism concentration in the tank and outlet PEF stream vary with time.

Assuming first order kinetics and plug flow, the mass balance for microorganisms in the PEF chamber gives the following equation (the same as that in

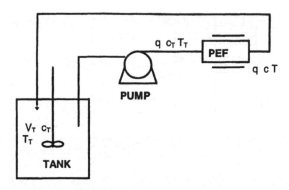

Figure 2.6 PEF recirculation mode.

the single pass operation):

$$c = c_T \exp(-kV_r/q) \tag{8}$$

Assuming perfect mixing in the tank, the non-stationary mass balance leads to:

$$c_T = c_{T0} \exp\{-q/V_T(1 - \exp[-kV_r/q])t\} \tag{9}$$

where c_{T0} (microorganisms/L) is the initial concentration of microorganisms, and V_T(L) is the tank volume. (V_r/q) and (V_T/q) are the residence times in PEF and tank, respectively.

Figure 2.7 shows the variation of the relative microorganism concentration (c_T/c_{T0}) in the tank and outlet PEF stream (c/c_{T0}) according to Equations (8) and (9). In Figures 2.7 and 2.8, the flow rate q is 0.5 L/min, the tank volume is 1.5 L, and the kinetic constant of the rate equation k is 1 s^{-1}.

Similar to the continuous operation, a part ϕ of the energy dissipated during the discharge of the capacitor Q (J/s) will increase the food temperature. In recirculation the energy balance in the PEF chamber is the same as that in the single pass operation:

$$Q\rho C_p(T - T_T) = \phi Q \tag{10}$$

Assuming non-stationary conditions, perfect mixing and adiabatic conditions (that gives the possible greater increase of temperature), the energy balance in the tank leads to the following linear relation between the temperature in the

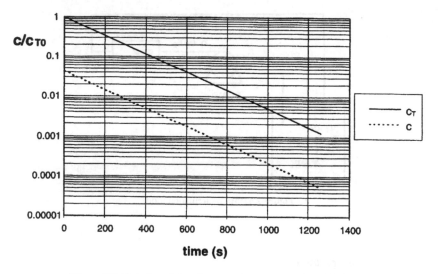

Figure 2.7 Variation of the microorganism concentration with time.

tank T, and time.

$$T_T = T_{T_0} + \frac{rQ}{V_T C_p \rho} t \tag{11}$$

where T_{T_0} is the initial tank temperature.

Figure 2.8 show this relationship in the case of one fluid with properties similar to water ($C_p = 4200$ J/(kg°C) and $\rho = 1000$ kg/M³) using a pulse

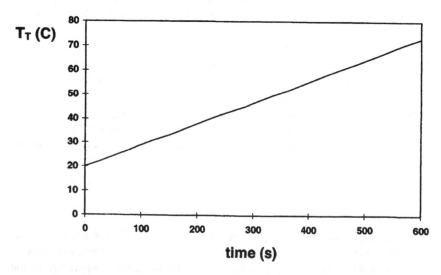

Figure 2.8 Variation of the tank temperature with time in the recirculation mode.

frequency of 4.3 Hz, a voltage of 36 kV, and a capacitor of 0.5 μF. The initial tank temperature was 20°C, and it was assumed that 40% of the energy dissipated during the capacitor discharge accumulates in the system ($\phi = 0.4$).

EXPERIMENTAL RESULTS

Suspensions containing *Bacillus subtilis* var. *niger*, ATCC 9372 were prepared in a standard simulated milk ultrafiltrate (SMUF). The culture was initially freeze-dried and allowed to grow in a nutrient broth of DIFCO for 3 to 4 hours. Afterwards, 2 mL was inoculated in 100 mL of the growth medium for 10 to 12 hours until the microorganisms reached stationary growth. Finally, the samples were centrifuged, buffered with phosphatase to pH 6.8, and mixed with 5 L (1/2 or 1/70) SMUF. Table 2.1 shows the characteristics of the SMUF used (Jenness and Koops, 1962).

SINGLE OPERATION

To avoid an excessive increase in the SMUF solution temperature, the flow rate used was $q = 0.5$ L/min and the operating voltage was $V = 36$ kV. The reactor volume was $V_r = 0.026$ L given a residence time of 3.1 s and a negligible increase in the temperature. Frequency pulses were selected in order to give the solution 5, 10, and 15 pulses. Two different SMUF/water ratios of 1/2 and 1/70 were used.

Table 2.2 and Figure 2.9 show the variation of the microorganism concentration in the outlet PEF stream with the pulse frequency obtained in the single pass operation with the above mentioned conditions.

The experimental results reveal the extensive influence of the SMUF/water ratio (different electric conductivity) on the elimination of microorganisms. With very diluted SMUF in water (1/70) and a total amount of 5 pulses, a

TABLE 2.1. Composition of SMUF (Simulated Milk Ultrafiltrate).

Components	Concentration (mg/L Water)
Lactose	50.00
Potassium phosphate monobasic	1.58
Tripotassium citrate	0.98
Trisodium citrate	1.79
Potassium sulfate	0.18
Calcium chloride dehydrate	1.32
Magnesium citrate	0.38
Potassium carbonate	0.30
Potassium chloride	1.08

TABLE 2.2. Experimental Results for a Single Pass Operation that Gives the Relative Microorganism Concentration in the Outlet PEF (c/c_T) for Different Conditions (Frequency Pulsing and SMUF/ Water Ratio).

	Ratio SMUF/Water	
f (Hz)	1/70	1/2
0	1	1
1.7	0.0001	0.68
3	0	0.44
5	0	0.0176

reduction in microbial population of four logarithmic cycles was obtained, whereas only a two logarithmic cycle reduction was reached using 15 pulses when the SMUF was concentrated (ratio SMUF/water = 1/2).

According to these results and Equation (8), the corresponding kinetic constants for a first order rate k were 2.95 s^{-1} for a SMUF/water ratio of 1/70 and 5 pulses, and 0.124 s^{-1}, 0.263 s^{-1}, and 1.29 s^{-1} for a SMUF/water ratio of 1/2 and 5, 10, and 15 pulses, respectively.

RECIRCULATION

In order to avoid thermal problems in the fluid food, while using the recycling mode the tank was surrounded by a mixture of ice and water to prevent heat transfer to the fluid food. Consequently, the operation was not carried out adiabatically.

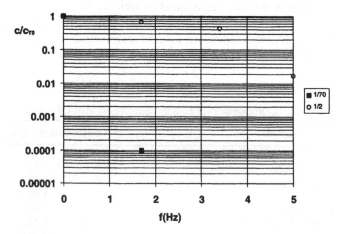

Figure 2.9 Experimental results of a single pass operation which gives the relative microorganism concentration in the outlet PEF ($c/c - r$) vs. the frequency pulsing for different SMUF/water ratio.

TABLE 2.3. Experimental Results of a Recycling Operation that Displays the Time Variation of the Relative Microorganism Concentration in a Tank (c_T/c_{T_0}) for Different SMUF/Water Ratios.

	Ratio SMUF/Water	
Time (s)	1/2	1/70
0	1	1
150	0.8	
300	0.52	0.257
450	0.32	
600	0.20	0.0129
900	0.100	0.00143
1200	0.0264	0.00143

Table 2.3 and Figure 2.10 show the variation of the relative microorganism concentration in the tank (c_T/c_{T_0}) with time, using a recirculation flow rate of 0.5 L/min for the fluid food and pulse conditions of 36 kV and 1.7 pulses/s. Similar to the single pass operation, two different ratios of SMUF/water were used. In both cases the initial concentration of microorganisms was 109 microorganisms/L.

Similar to the single pass operation, when the SMUF was more diluted the removal of microorganisms was higher.

Figures 2.11 and 2.12 compare the experimental results obtained and the predictions for a single pass operation. Theoretical lines in both figures correspond

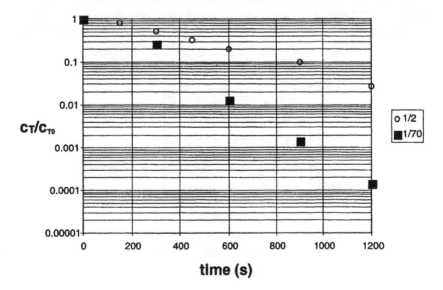

Figure 2.10 Experimental results in a recirculation mode.

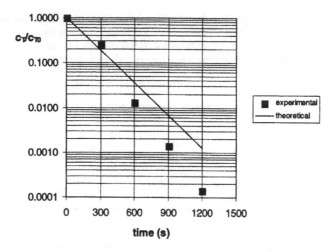

Figure 2.11 Experimental and expected results for a SMUF/water ratio of 1/70.

to the plot of Equation (9) (mathematical model for recycling operation) using the corresponding kinetic k-values obtained in the experiments for the continuous operation. Because the pulse frequency used was 1.7 pulses/s, the k-value used for the SMUF/water ratio of 1/2 was 0.123 s^{-1}, and 2.95 s^{-1} for the SMUF/water ratio of 1/70.

Taking into account the accuracy obtained in the determination of microorganism concentration, the difference between the experimental results in recirculation mode and the expected results from the data obtained in the single pass

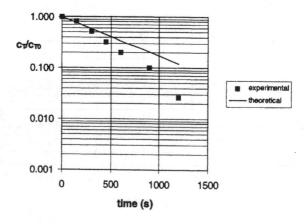

Figure 2.12 Experimental and expected results for a SMUF/water ratio of 1/2.

operation was good. In addition, the flow regime in the chamber was laminar during the experiments (Reynolds number 120), but the approximation of the plug flow model for the fluid was not as good. However, the plug flow model gave very simple equations, that in this case had a good fit with the experimental results. It should also be noted that in both cases the experimental results were better than the model prediction.

CONCLUSIONS

According to the experimental results, the following conclusions may be drawn:

- PEF technology is very promising for the nonthermal inactivation of microorganisms.
- According to the mathematical model, when the voltage or conductivity of a fluid food increases, the temperature in the outlet stream of PEF also increases. It is thus necessary to remove more heat from the fluid food in order to maintain a nonthermal treatment.
- The experimental results obtained in a single pass operation mode may give a good approximation of the results to be obtained in the recirculation mode.

NOMENCLATURE

C = capacitor (F)
c = microorganism concentration in the outlet PEF stream (microorganisms/L)
C_p = specific heat [J/(kg°C)]
c_T = microorganism concentration in the tank (microorganisms/L)
c_{T_0} = initial microorganism concentration in the tank (microorganisms/L)
E = energy dissipated by the capacitor (J)
F = pulse frequency (L/s)
k = first order kinetic constant (L/s)
Q = heat flow rate (J/s)
q = volumetric flow rate (L/s)
R_s = resistance (Ω)
T = temperature in the outlet PEF stream (°C)
t = time (s)
T_T = tank temperature (°C)
T_{T_0} = initial tank temperature (°C)
V = voltage (V)
V_r = reactor volume (L)

V_T = tank volume (L)
ϕ = r = part of the energy released to the fluid
ρ = fluid density (kg/L)

ACKNOWLEDGEMENT

The authors wish to thank the Generalitat de Catalunya (CIRIT) for supporting the travel and lodging expenses of Santiago Esplugas.

REFERENCES

Barbosa-Cánovas, G. V., Swanson, B. G., and Pothakamury, U. 1995. State of the art technologies for the stabilization of foods by nonthermal processes: Physical methods. In: *Preservation of Foods by Moisture Control. Fundamentals and Applications*. Edited by G. V. Barbosa-Cánovas and J. Welti-Chanes. Technomic Publishing Co., Inc., Lancaster, PA. 493–532.

Hülsherger, H., Pottel, J., and Niemann, E. G. 1983. Electric field effects on bacteria and yeast cells. *Radiat. Environ. Biophys.* 22:149–162.

Jenness, R. and Koops, J. 1962. Preparation and properties of a salt solution which simulates milk ultrafiltrate. *Neth. Milk Dairy J.* 16(3):153–164.

Kinosita, K. Jr. and Tsong , T. Y. 1977. Voltage induced pore formation and haemolysis erythrocites. *Biochim. Biophys. Acta* 471:227–242.

Levenspiel, O. 1972. *Chemical Reaction Engineering*. Wiley, New York.

Qin, B. L., Zhang, Q., Barbosa-Cánovas, G. V., Swanson, B. G., and Pedrow, P. D. 1995. Pulsed electric field treatment chamber design for liquid food pasteurization using a finite element method. *Transactions of the ASAE* 38 (2):557–565.

Sale, A. J. H. and Hamilton, W. A. 1967. Effects of high electric fields on microorganisms. I Killing of bacteria and yeast. *Biochim. Biophys. Acta* 148:781–788.

Sale, A. J. H. and Hamilton, W. A. 1968. Effects of high electric fields on microorganisms. III Lysis of erythrocytes and protoplasts. *Biochim. Biophys. Acta* 163:34–43.

Zhang, Q., Barbosa-Cánovas, G., and Swanson, B. 1995. Engineering aspects of pulsed electric field pasteurization. *Journal of Food Engineering* 25:261–281.

Zimmermann, U., Vienken, J., and Pilwat, G. 1980 Development of a drug carrier system: Electric field induced effects in cell membranes. *Bioelectrochem. Bioenerg.* 7:553–574.

CHAPTER 3

Physical Properties of Liquid Foods for Pulsed Electric Field Treatment

K. T. RUHLMAN
Z. T. JIN
Q. H. ZHANG

ABSTRACT

THE effectiveness of pulsed electric field (PEF) processing for a liquid product depends on the product's physical properties, including electrical conductivity, density, and viscosity. These properties of liquid products were measured at temperatures ranging from 4°C to 60°C, and used to calculate change in temperature (ΔT), power-input (P), and Reynolds number (N_{Re}). The results showed that with an increase in temperature there was an increase in conductivity, ΔT, P, and N_{Re} and a decrease in density and viscosity.

INTRODUCTION

Pulsed electric field (PEF) processing is a nonthermal method used to increase shelf life and maintain food safety by inactivating spoilage and pathogenic microorganisms. Many researchers have demonstrated this effect including Sale and Hamilton (1967a, 1967b), Mizuno and Hori (1988), Jayaram et al. (1992), Qin et al. (1994), and Pothakmury et al. (1996). This method of processing is advantageous because the change in product color, flavor, and nutritive value during processing is minimized (Dunn and Pearlman, 1987; Jin and Zhang, 1999; Jin et al., 1998, Jia et al., 1999).

To process the food with PEF in a continuous system, the fluid flows through a series of treatment zones with a high voltage electrode on one side of each zone and a low voltage electrode on the other. The PEF process is defined by the

45

electric field strength, or the voltage applied per distance between electrodes and treatment time.

To design and optimize the operation of PEF processing units, information on the physical properties of the products being processed is required over a wide range of temperatures. The physical properties that are most critical are electrical conductivity, density, specific heat, and viscosity. Published data on these physical properties of specific liquid foods are limited.

Liquid foods contain several ion species that carry an electrical charge and allow them to conduct electricity. At a given voltage, the electrical current flow is directly proportional to the electrical conductivity of the food (Zhang et al., 1995). An increase in electrical conductivity causes an increase in the overall energy input and change in temperature during processing at a defined dosage.

The density and specific heat of the food affects the amount of temperature change that occurs during PEF treatment. As the density of the product decreases, the total temperature change increases (Zhang et al., 1995). Similarly, a decrease in specific heat also increases the temperature change during processing.

The viscosity of the product determines the flow characteristics, which are calculated by the Reynolds number. For a Reynolds number greater than 2100, the flow is turbulent, which provides a uniform velocity profile. A uniform velocity profile in the treatment chamber is likely to provide a uniform PEF process.

The objective of this study was to determine the electrical conductivity, density, and viscosity of liquid foods to provide a database of physical properties to those who wish to design or operate a PEF treatment process.

MATERIALS AND METHODS

Samples of different types of coffee (3), beer (2), fruit juice (6), milk (6), and vegetable juice (3) were obtained from a local grocery store. Samples were selected to represent liquid foods of different compositions. For each of the samples, the electrical conductivity, density, and apparent viscosity were measured a minimum of three times at 4°C, 22°C, 30°C, 40°C, 50°C, and 60°C.

ELECTRICAL CONDUCTIVITY

The electrical conductivity was measured using a Yellow Springs Instrument Company conductivity meter (YSI model 30 Handheld Salinity, Conductivity, and Temperature System, YSI Incorporated, Yellow Springs, OH). The sample was placed into a 500 mL bottle and the bottle was set into a water bath at a

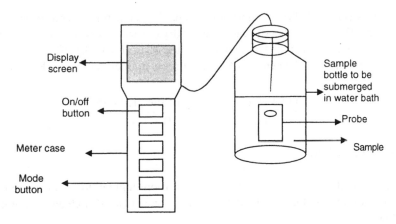

Figure 3.1 Measurement of electrical conductivity using a YSI electrical conductivity meter.

determined temperature (Figure 3.1). The electrical conductivity probe was completely submerged in the liquid sample. The probe measured both the electrical conductivity and the temperature of the sample. Once the desired temperature was achieved, the value for the electrical conductivity was recorded.

DENSITY

A determination of density was made using a hydrostatic method based on the measurement of the buoyant force acting on a tare body submerged in the liquid product (Constenla et al., 1989). A 250 mL beaker filled with the sample was placed into a water bath at a determined temperature. Once the desired temperature was achieved, a weighted body was submerged in the sample and weighed using a Mettler balance (Figure 3.2). The weight of the body submerged in the

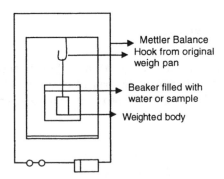

Figure 3.2 Measurement of density using a hydrostatic method.

sample was recorded (s). Calibration was performed using water at room temperature. The weighted body was weighed dry (d) and submerged in water (w). The density of the sample was then calculated according to the following equation, assuming that the density of water at room temperature was 1000 kg/m^3.

$$\frac{1000\,\text{kg/m}^3}{(d) - (w)} = \frac{\text{density of sample kg/m}^3}{(d) - (s)}$$

SPECIFIC HEAT

The specific heat (C_p) was calculated at room temperature using a model estimation for food materials of high water content (w) (Singh and Heldman, 1993).

$$C_p = 1.675 + 0.025\,w$$

The water content for each sample was determined by drying the sample in a vacuum oven.

APPARENT VISCOSITY

The kinematic viscosity was measured using a Cannon-Fenske capillary tube viscometer (Fisherbrand kinematic viscometer tubes, ASTM Sizes 50, 100, and 200, Fisher, Pittsburgh, PA), calibrated at specific temperatures with de-ionized water (Constenla et al., 1989). The efflux time, or the time it takes for the sample to flow through the viscometer, was measured at least three times using at least two different viscometers. The kinematic viscosity was calculated by multiplying the efflux time viscometer constant. The apparent viscosity was calculated for each product by multiplying the kinematic viscosity by the density.

PROCESS CALCULATIONS

The total possible temperature change per pair of treatment chambers (ΔT), total energy input during treatment per pair of chambers (P), and Reynolds number (N_{Re}) were calculated for all products at all temperatures using the following equations:

$$\Delta T = (E^2 t\sigma / \rho\, C_p)/n$$
$$P = E^2 t\sigma / n$$
$$N_{Re} = \rho\, Du / \mu$$

The variables in these equations are electrical conductivity, density, viscosity, and specific heat, which are specific for each product at a given temperature. For the calculations, the PEF system designed for processing orange juice was used as an example (Qiu et al., 1998), where:

E = electric field strength (3.2×10^6 V/m)
t = total treatment time (9.0×10^7 s)
n = number of pairs of treatment chambers (6)
D = treatment zone diameter (4.8×10^{-3} m)
μ = mean liquid velocity (1.3 m/s)
σ = product electrical conductivity (S/m)
ρ = product density (kg/m^3)
C_p = product specific heat (kJ/kg°C)

RESULTS AND DISCUSSION

ELECTRICAL CONDUCTIVITY

Electrical conductivity increased with an increase in temperature (Table 3.1). The variability due to the change in temperature was in the range of 55% to 67% for all products tested. The vegetable juice samples had the greatest overall electrical conductivity compared to the other products, while the beer samples had the lowest. The vegetable juice was higher due to the presence of salt in the formulation. Ionic species present in the food, such as salts and acids, act as electrolytes that allow an electric current to pass through the food (Halden et al., 1990). The fruit juice group showed the greatest variability between products (79%), while the milk group showed a very low variability between products (21%). The typical electrical conductivity of normal milk is between .40 and .55 S/m at 25°C (Nielen et al., 1992). All of the experimental measurements for the milk products at 22°C fall within this range, regardless of fat content or presence of chocolate. The large variation in the fruit juice products can be explained by the difference in the physical properties and ionic make up of the original fruits.

The increase in energy input (Figure 3.3) and change in temperature (Figure 3.4) are directly proportional to the increase in electrical conductivity. As temperature increases, due to the energy transfer during treatment, the conductivity of the food and the energy input needed to achieve a dosage level increase. There is a dynamic relationship between electrical conductivity, energy input, and temperature increase.

According to Qiu et al. (1998), a 99.9% reduction of the natural microflora in orange juice was achieved using a 35 kV/cm electric field strength and a treatment time of 60 μs. Of the fruit juice samples, the orange juice had the

TABLE 3.1. The Electrical Conductivity (S/M)* of Liquid Products Measured at Increasing Temperatures.

	4°C	22°C	30°C	40°C	50°C	60°C
Beer						
Beer	0.080 ± .013	0.143 ± .010	0.160 ± .009	0.188 ± .003	0.227 ± .013	0.257 ± .006
Light beer	0.083 ± .009	0.122 ± .003	0.143 ± .003	0.167 ± .006	0.193 ± .008	0.218 ± .008
Coffee						
Black coffee	0.138 ± .003	0.182 ± .008	0.207 ± .011	0.237 ± .012	0.275 ± .013	0.312 ± .024
Coffee with milk	0.265 ± .013	0.357 ± .020	0.402 ± .018	0.470 ± .026	0.550 ± .050	0.633 ± .058
Coffee with sugar	0.133 ± .003	0.185 ± .009	0.210 ± .013	0.250 ± .009	0.287 ± .013	0.323 ± .008
Fruit Juice						
Apple juice	0.196 ± .014	0.239 ± .020	0.279 ± .038	0.333 ± .040	0.383 ± .049	0.439 ± .052
Cranberry juice	0.063 ± .006	0.090 ± .005	0.105 ± .005	0.123 ± .003	0.148 ± .013	0.171 ± .009
Grape juice	0.056 ± .010	0.083 ± .013	0.092 ± .009	0.104 ± .004	0.122 ± .006	0.144 ± .007
Lemonade	0.084 ± .014	0.123 ± .004	0.143 ± .004	0.172 ± .003	0.199 ± .001	0.227 ± .006
Limeade	0.090 ± .005	0.117 ± .008	0.137 ± .003	0.163 ± .006	0.188 ± .003	0.217 ± .006
Orange juice	0.314 ± .032	0.360 ± .009	0.429 ± .016	0.500 ± .022	0.600 ± .002	0.690 ± .010
Milk						
Chocolate 3% fat milk	0.332 ± .020	0.433 ± .036	0.483 ± .015	0.567 ± .029	0.700 ± .050	0.800 ± .001
Chocolate 2% fat milk	0.420 ± .040	0.508 ± .014	0.617 ± .029	0.700 ± .050	0.833 ± .076	1.000 ± .044
Chocolate skim milk	0.532 ± .011	0.558 ± .017	0.663 ± .016	0.746 ± .009	0.948 ± .004	1.089 ± .001
Lactose free milk	0.380 ± .013	0.497 ± .006	0.583 ± .029	0.717 ± .029	0.817 ± .029	0.883 ± .029
Skim milk	0.328 ± .002	0.511 ± .001	0.599 ± .004	0.713 ± .001	0.832 ± .001	0.973 ± .003
Whole milk	0.357 ± .016	0.527 ± .064	0.617 ± .076	0.683 ± .029	0.800 ± .004	0.883 ± .029
Vegetable Juice						
Carrot juice	0.788 ± .035	1.147 ± .041	1.282 ± .072	1.484 ± .095	1.741 ± .023	1.980 ± .058
Tomato juice	1.190 ± .012	1.697 ± .023	1.974 ± .019	2.371 ± .008	2.754 ± .013	3.140 ± .004
Vegetable juice cocktail	1.087 ± .004	1.556 ± .014	1.812 ± .001	2.141 ± .004	2.520 ± .012	2.828 ± .016

*Values are an average of at least three measurements.

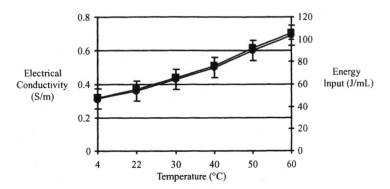

Figure 3.3 Electrical conductivity and calculated temperature change per pair of chambers vs. input temperature for orange juice. Error bars represent ±1 standard deviation.

greatest overall electrical conductivity (Table 3.1), and therefore the greatest energy input (Figure 3.5). To process any of the other fruit juice types, less power will be needed to achieve the same electric field strength and microbial inactivation.

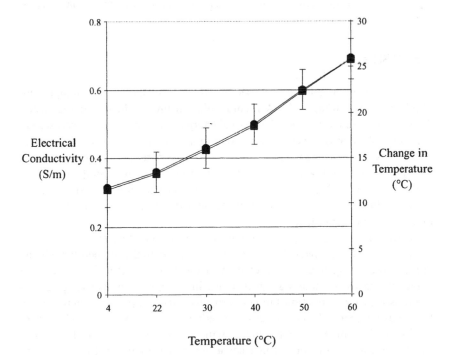

Figure 3.4 Electrical conductivity and calculated temperature change per pair of chambers vs. input temperature for orange juice. Error bars represent ±1 standard deviation.

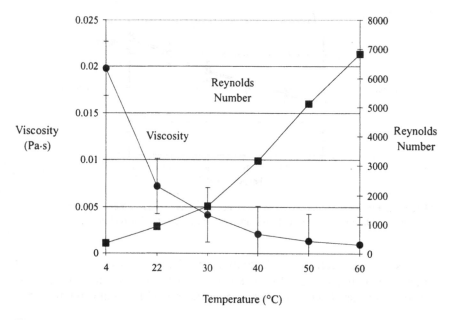

Figure 3.5 Viscosity and Reynolds number vs. temperature for chocolate 3% fat milk. Error bars represent ±1 standard deviation.

VISCOSITY

The viscosity of all products decreased with an increase in temperature (Table 3.2). According to the Arrhenius relationship, as the product is heated, the viscosity decreases, since the thermal energy of the molecules increases and the intermolecular distances increase due to the thermal expansion (Constenla et al., 1989). The difference in viscosity due to temperature was in the range of 69% to 77% for all products except the milk group. The milk products differed in the range of 80% to 95% with the increase in temperature. Milk contains several different molecules, including fat and protein, that are not present in many of the other products.

Of the products tested, milk products have the greatest overall viscosity, while coffee and beer have the lowest (Table 3.2). With the temperature change during processing, there was a change in the product viscosity. With a decrease in viscosity, there was an increase in Reynolds number (Figure 3.5). There is a dynamic relationship between viscosity, temperature increase, and Reynolds number. When the viscosity is reduced to approximately 0.006 Pa·s, the Reynolds number of the product flowing at 85 L/hr exceeded 2100. Above 2100, the flow is turbulent, which has a uniform velocity profile. A uniform velocity profile provides a uniform PEF process.

TABLE 3.2. The Viscosity (Pa•S)* of Liquid Products Measured at Increasing Temperatures.

	4°C	22°C	30°C	40°C	50°C	60°C
Beer						
Beer	0.002688 ± .000210	0.001376 ± .000112	0.001079 ± .000090	0.000830 ± .000247	0.000669 ± .000053	0.000603 ± .000011
Light beer	0.002305 ± .000211	0.001238 ± .000093	0.000986 ± .000067	0.000784 ± .000050	0.000639 ± .000034	0.000574 ± .000009
Coffee						
Black coffee	0.001616 ± .000100	0.000977 ± .000028	0.000817 ± .000022	0.000664 ± .000010	0.000558 ± .000017	0.000502 ± .000032
Coffee with sugar	0.001922 ± .000159	0.001121 ± .000095	0.000917 ± .000058	0.000734 ± .000074	0.000614 ± .000042	0.000567 ± .000015
Coffee with milk	0.002318 ± .000167	0.001301 ± .000091	0.001064 ± .000081	0.000842 ± .000054	0.000686 ± .000053	0.000611 ± .000011
Fruit Juice						
Apple juice	0.002445 ± .000218	0.001372 ± .000125	0.001115 ± .000106	0.000869 ± .000067	0.000718 ± .000067	0.000627 ± .000019
Cranberry juice	0.002723 ± .000378	0.001475 ± .000175	0.001182 ± .000130	0.000925 ± .000103	0.000746 ± .000073	0.000653 ± .000041
Grape juice	0.002456 ± .000193	0.001350 ± .000107	0.001114 ± .000105	0.000875 ± .000081	0.000710 ± .000065	0.000632 ± .000009
Lemonade	0.002416 ± .000032	0.001338 ± .000012	0.001075 ± .000008	0.000839 ± .000007	0.000669 ± .000038	0.000642 ± .000008
Limeade**	0.002879 ± .000235	0.001540 ± .000073	0.001222 ± .000025	0.000933 ± .000023	0.000764 ± .000018	0.000684 ± .000029
Orange juice**	0.004398 ± .000038	0.002407 ± .000224	0.001847 ± .000160	0.001355 ± .000104	0.001039 ± .000064	0.000922 ± .000043
Milk						
Chocolate 3% fat milk	0.019791 ± .000156	0.007173 ± .000203	0.004113 ± .000082	0.002081 ± .000025	0.001282 ± .000031	0.000959 ± .000017
Chocolate 2% fat milk	0.013209 ± .000367	0.005262 ± .000123	0.003230 ± .000085	0.001616 ± .000111	0.001040 ± .000009	0.000872 ± .000020
Chocolate skim milk	0.004012 ± .000078	0.002070 ± .000043	0.001533 ± .000012	0.001175 ± .000011	0.000940 ± .000005	0.000824 ± .000018
Lactose free milk	0.003742 ± .000046	0.001725 ± .000050	0.001320 ± .000006	0.000992 ± .000006	0.000773 ± .000005	0.000703 ± .000005
Whole milk	0.003423 ± .000054	0.001605 ± .000015	0.001215 ± .000004	0.000937 ± .000013	0.000717 ± .000005	0.000670 ± .000002
Skim milk	0.003901 ± .000475	0.001881 ± .000199	0.001422 ± .000130	0.001094 ± .000084	0.000875 ± .000056	0.000767 ± .000083
Vegetable Juice						
Carrot juice**	0.003360 ± .000410	0.001846 ± .000178	0.001504 ± .000123	0.001188 ± .000087	0.000943 ± .000054	0.000829 ± .000092
Tomato juice**	0.003163 ± .000045	0.001755 ± .000135	0.001395 ± .000010	0.001068 ± .000011	0.000865 ± .000008	0.000797 ± .000005
Vegetable juice cocktail**	0.003302 ± .000041	0.001664 ± .000008	0.001316 ± .000005	0.001051 ± .000008	0.000834 ± .000008	0.000784 ± .000009

*Values are an average of at least three measurements.
**Juices were clarified before measurement.

TABLE 3.3. The Density (kg/M³)* of Liquid Products Measured at Increasing Temperatures.

	4°C	22°C	30°C	40°C	50°C	60°C
Beer						
Beer	1040.00 ± 2.76	1020.00 ± 2.31	1000.00 ± 3.72	1000.00 ± 2.31	1000.00 ± 1.02	990.00 ± 2.04
Light beer	1013.83 ± 0.11	1012.23 ± 1.48	1005.27 ± 0.01	1002.54 ± 1.04	999.83 ± 0.19	997.50 ± 3.49
Coffee						
Black coffee	1004.92 ± 1.64	1003.10 ± 2.17	1001.88 ± 2.49	999.80 ± 0.95	995.98 ± 2.51	991.88 ± 2.56
Coffee with sugar	1020.65 ± 0.03	1018.17 ± 0.08	1016.70 ± 0.16	1014.75 ± 0.05	1010.43 ± 0.03	1006.78 ± 0.30
Coffee with milk	1035.53 ± 0.95	1033.38 ± 0.69	1031.83 ± 1.34	1028.73 ± 0.21	1026.92 ± 2.63	1021.62 ± 0.04
Fruit Juice						
Apple juice	1058.06 ± 9.50	1054.54 ± 8.90	1052.86 ± 10.1	1049.42 ± 9.00	1046.83 ± 10.8	1043.11 ± 8.00
Cranberry juice	1061.74 ± 8.10	1057.97 ± 7.40	1055.99 ± 7.00	1052.32 ± 6.60	1048.84 ± 5.90	1045.03 ± 6.40
Grape juice	1055.54 ± 1.66	1052.35 ± 2.79	1050.04 ± 1.84	1047.13 ± 2.24	1044.19 ± 1.53	1040.97 ± 2.80
Lemonade	1065.90 ± 0.30	1062.87 ± 0.97	1060.00 ± 0.34	1057.04 ± 0.38	1054.38 ± 0.01	1050.84 ± 0.54
Limeade	1064.12 ± 8.20	1061.45 ± 7.14	1058.97 ± 5.47	1054.65 ± 5.27	1051.07 ± 6.13	1047.43 ± 5.41
Orange juice	1056.86 ± 3.40	1053.57 ± 3.80	1051.82 ± 3.40	1048.80 ± 3.40	1045.75 ± 3.60	1041.74 ± 3.70
Milk						
Chocolate 3% fat milk	1066.32 ± 2.04	1058.69 ± 0.48	1056.42 ± 0.07	1051.23 ± 0.22	1046.89 ± 0.39	1045.27 ± 0.05
Chocolate 2% fat milk	1060.67 ± 0.98	1056.47 ± 0.10	1053.21 ± 0.04	1048.89 ± 0.13	1044.99 ± 0.32	1040.94 ± 0.73
Chocolate skim milk	1065.81 ± 1.69	1060.77 ± 1.46	1057.64 ± 1.51	1053.07 ± 3.05	1049.33 ± 4.00	1042.95 ± 4.92
Lactose free milk	1039.30 ± 0.57	1033.86 ± 0.27	1030.06 ± 0.36	1026.17 ± 0.07	1023.20 ± 0.58	1018.55 ± 0.83
Skim milk	1033.71 ± 3.75	1031.39 ± 4.26	1028.28 ± 4.14	1025.69 ± 4.38	1023.01 ± 4.33	1020.04 ± 4.04
Whole milk	1034.70 ± 0.16	1029.56 ± 0.51	1027.20 ± 1.70	1023.01 ± 0.40	1018.69 ± 1.05	1015.86 ± 1.89
Vegetable Juice						
Carrot juice	1036.88 ± 3.28	1033.22 ± 3.75	1031.24 ± 4.02	1028.32 ± 3.78	1025.58 ± 4.00	1021.09 ± 3.81
Tomato juice	1024.02 ± 2.61	1020.96 ± 3.88	1019.59 ± 3.73	1015.61 ± 4.25	1012.04 ± 4.27	1008.86 ± 4.02
Vegetable juice cocktail	1023.81 ± 4.00	1019.99 ± 4.02	1017.24 ± 4.06	1014.94 ± 3.99	1011.96 ± 4.32	1008.36 ± 4.01

*Values are an average of at least three measurements.

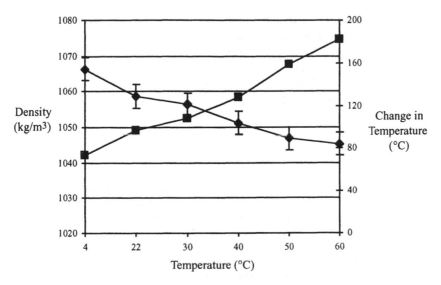

Figure 3.6 Density and calculated temperature change per pair of chambers vs. temperature for chocolate 3% fat milk. Error bars represent ±1 standard deviation.

DENSITY

With an increase in temperature, there is a decrease in the measured values for density (Table 3.3). The density of the products depends on the intermolecular forces and water solute interactions, which are affected by temperature (Constenla et al., 1989). The calculated difference in density due to the temperature increase from 4°C to 60°C is in the range of 1.3% to 1.6% for all products except milk. The milk products has a variation in the range of 1.8–2.0%. Again, this difference in density between milk and the other groups of products is due to the presence of the large fat and protein molecules. With the decrease in product density, there was a corresponding increase in the temperature change during processing (Figure 3.6).

CONCLUSIONS

Products that have low electrical conductivity and viscosity, and high density, will be the easiest and most energy efficient to process using PEF. By using a cooling heat exchanger between each pair of treatment chambers, the temperature increase can be kept to a minimum. By maintaining a specific temperature, the process can be maintained at a specific energy input to achieve a desired dosage.

The efficiency of PEF treatment for a specific product depends on the physical properties of the product, including electrical conductivity, viscosity, and density. The importance of temperature control in relation to PEF treatment has been illustrated. A change in one physical property could alter the entire PEF process of the food. This database provides critical information to those who design and operate a pulsed electric field treatment process.

REFERENCES

Constenla, D. T., Lozano, J. E., and Crapiste, G. H. 1989. Thermophysical properties of clarified apple juice as a function of concentration and temperature. *J. Food Sci.* 54(3), 663–668.

Dunn, J. E. and Pearlman, J. S. 1987. Methods and apparatus of extending the shelf-life of fluid food products. U.S. Patent 4,695,472.

Halden, K., DeAlwis, A. A. P., and Fryer, P. J. 1990. Changes in the electrical conductivity of foods during ohmic heating. *Int. J. of Food Sci. and Tech.* 25, 9–25.

Jayaram, S., Castle, G. S. P., and Margaritis, A. 1992. Kinetics of sterilization of *Lactobacillus brevis* cells by the application of high voltage pulses. *Biotech. and Bioeng.* 40, 1412–1420.

Jia, M., Zhang, Q. H., and Min, D. B. 1999. Pulsed electric field processing effects on flavor compounds and microorganisms of orange juice. *Food Chemistry.* 65(4), 445.

Jin, Z. T. and Zhang, Q. H. 1999. Pulsed electric field treatment inactivates microorganisms and preserves quality of cranberry juice. *J. Food Proc. Pres.* 23(6), 481–499.

Jin, Z. T., Ruhlman, K. T., Qiu, X., Jia, M., Zhang, S., and Zhang, Q. H. 1998. Shelf life evaluation of pulsed electric fields treated aseptically packaged cranberry juice. *98 IFT Annual Meeting.* Atlanta, GA. June 20–24. Book of Abstracts p. 70.

Mizuno, A. and Hori, Y. 1988. Destruction of living cells by pulsed high voltage application. *IEEE Transactions on Industry Applications.* 24(3), 387–394.

Nielen, M., Deluyker, H., Schukken, Y. H., and Brand, A. 1992. Electrical conductivity of milk: Measurement, modifiers, and meta analysis of mastitis detection performance. *J. Dairy Sci.* 75, 606–614.

Pothakamury, U. R., Vega-Mercado, H., Zhang, Q. H., Barbosa-Cánovas, G. V., and Swanson, B. G. 1996. Effect of growth stage and processing temperature on the inactivation of *E. coli* by pulsed electric fields. *J. Food Prot.* 59(11), 1167–1171.

Qin, B., Zhang, Q. H., Barbosa-Cánovas, G. V., Swanson, B. G., and Pedrow, P. D. 1994. Inactivation of microorganisms by pulsed electric fields of different voltage waveforms. *IEEE Transactions on Dielectrics and Electrical Insulation.* 1(6), 1047–1057.

Qiu, X., Sharma, S., Tuhela, L., and Zhang, Q. H. 1998. An integrated PEF pilot plant for continuous nonthermal pasteurization of fresh orange juice. *Trans. ASAE.* 41(4), 1069–1074.

Sale, A. J. H. and Hamilton, W. A. 1967a. Effects of high electric fields on microorganisms I. Killing of bacteria and yeasts. *Biochim. Biophys. Acta.* 148(3), 781–788.

Sale, A. J. H. and Hamilton, W. A. 1967b. Effects of high electric fields on microorganisms II. Mechanism of action of the lethal effect. *Biochim. Biophys. Acta.* 148(3), 789–800.

Singh, R. P. and Heldman, D. R. 1993. *Introduction to Food Engineering.* Academic Press Inc., San Diego.

Zhang, Q. H., Barbosa-Cánovas, G. V., and Swanson, B. G. 1995. Engineering aspects of pulsed electric field pasteurization. *J. Food Eng.* 25, 261–291.

CHAPTER 4

Enzymatic Inactivation by Pulsed Electric Fields: A Review

H. W. YEOM
Q. H. ZHANG

HIGH voltage pulsed electric field (PEF) systems have been studied as one of the most promising nonthermal food preservation methods. In terms of the microbial inactivation mechanism by PEF, extensive research has been done and the membrane rupture theory is now commonly accepted. However, for the inactivation effect of PEF on enzymes, there are very few reports, and it is still a controversial subject. The comparison of PEF research requires caution in considering different PEF systems that may affect enzyme activity. Generally speaking, a PEF system consists of several parts including a high voltage and pulse generator, a treatment chamber, and a fluid handling system (Table 4.1). The enzyme suspension medium is also important due to the effect this medium may have on enzyme activity under a high voltage pulsed electric field.

Gilliland and Speck (1967) found that lactic dehydrogenase, trypsin, and proteinases of *Bacillus subtilis* were inactivated by electrohydraulic discharge, a submerged high voltage spark system. No residual activity was observed in lactate dehydrogenase solution after 10 discharges at 10 kV using an electrode gap of 0.125 in. Trypsin was also significantly inactivated after 40 discharges. Protease activity of *B. subtilis* cells showed a 60% reduction after 40 discharges. Electrohydraulic treated buffer had no effect on the enzyme activity. Oxidation of free sulfhydryl groups and ferrous sulfate was observed. Oxidation of key components, based on the formation of free radicals in the electrohydraulic discharge system, was proposed as the enzyme inactivation mechanism.

Hamilton and Sale (1967) measured several enzyme activities in cell extracts after d.c.-pulse treatment of intact cell suspensions. The field strength was limited to less than 30 kV/cm by the electrical breakdown of air above the sample. NADH dehydrogenase, succinic dehydrogenase, and hexokinase of

57

TABLE 4.1. **Pulsed Electric Field System.**

High voltage and pulse generator	• Electric field strength (kV/cm) • Pulse duration time (μsec or msec) • Frequency (Hz) • Wave form (square or exponential decay)
Treatment chamber	• Electrode material • Electrode gap distance • Electrode configuration • Air contact or not
Fluid handling system	• Continuous or batch process • Flow rate (mL/s) • Cooling system (temperature control)
Enzyme suspension medium	• Water • Buffer • Simulated food • Natural food

E. coli 8196 were not significantly inhibited after d.c.-pulse treatment of the cells at 20–25 kV/cm. The acetylcholinesterase activity of bovine erythrocytes was not reduced after pulse treatment of the erythrocytes. Cellular compartments such as the cell wall might function as a barrier protecting intracellular enzymes against the electric field. Activities of lipase and α-amylase treated at voltages up to 30 kV/cm were not inhibited (data not shown).

Castro (1994) observed a 65% reduction of alkaline phosphatase activity in simulated milk ultrafiltrate at an electric field strength of 22 kV/cm using the static chamber of a commercial electroporation system. The temperature of the milk in the cuvette was measured by inserting a thermocouple immediately after 70 pulses. The temperature increased to 44°C due to Joule heating. Polyacrylamide gel electrophoresis of alkaline phosphatase treated with 70 pulses at 20.8 kV/cm revealed a single protein band identical to the untreated protein, demonstrating that PEF does not cause the hydrolysis of alkaline phosphatase. Polarization of alkaline phosphatase leading to aggregation of the enzyme was proposed as the enzyme inactivation mechanism of PEF.

Vega et al. (1995a) reported a 90% reduction of plasmin (milk alkaline protease) activity in a simulated milk ultrafiltrate at 30 and 45 kV/cm after 50 pulses. The sample temperature was controlled by using heat exchangers on the electrodes, and measured by using a thermocouple attached to the surface of the heat exchanger. Processing temperature was maintained at 15°C, demonstrating that the inactivation of plasmin with a pulsed electric field was nonthermal. The authors suggested that there is a synergistic effect between selected factors (e.g., the number of pulses, electric field, and temperature) in the inactivation of plasmin. The activity of PEF-treated plasmin was not restored after 24 hr of storage at 4°C. This result indicated that permanent inactivation of plasmin could be achieved by PEF. The inactivation mechanism of plasmin was explained by the changes in protein charge and configuration due to the electrostatic nature of plasmin. Vega et al. (1995b) also applied a pulsed electric field of 20 and 35 kV/cm to protease from *Pseudomonas fluorescens* using a pilot plant pulser. Approximately a 70% reduction of protease activity was observed at 35 kV/cm after 20 pulses. The authors found that the reduction of enzyme activity was related to both electric field strength and number of pulses.

Grahl and Markl (1996) studied the effect of PEF on raw milk enzymes such as lipase, lactoperoxidase, and alkaline phosphatase. At 21.5 kV/cm and high energy input ($Q = 400$ kJ/L), a 65% reduction of lipase activity and a 25% reduction of peroxidase activity were observed. The effect of PEF on alkaline phosphatase in raw milk was negligible (less than 5% reduction). The higher fat content of the milk gave a better protection effect against electrical field pulses.

Ho et al. (1997) reported the effects of high voltage pulses on activities of eight enzymes. Lipase, glucose oxidase, and heat stable α-amylase exhibited a large reduction in activity from 75% to 85%; peroxidase and polyphenol oxidase also showed a 30% to 40% reduction. However, alkaline phosphatase showed only a 5% reduction under high voltage pulses. Lysozyme and pepsin were activated at around 50 kV/cm. No change in temperature or pH was observed, indicating that any change in enzyme activity was due to electric field pulses. The authors proposed that the secondary and tertiary structure of the enzyme might play an important role in the inactivation of enzymes based on the results showing different degrees of inactivation according to the enzyme type. In comparison with Castro's results (1994) regarding the inactivation of alkaline phosphatase, the authors suggested that pulse width and pulse waveform may be more important than the electric field strength in the reduction of enzyme activity.

Yeom et al. (1999) conducted experiments with papain protease suspended in a 1 mM EDTA solution. After reaction with reducing agents, activated papain was treated with a pulsed electric field in a continuous system with up to 50 kV/cm of electric field strength at 10°C. A linear relation between residual activity and electric field strength was observed. Further inactivation of papain, separated from activators, was observed within 24 hours of storage after PEF treatment. No significant thermal or chemical effect was found. The loss of α-helical structure, which is related to the papain activity, was observed. The protein structural change by PEF was proposed as one of enzymatic inactivation mechanism.

SUMMARY

As shown in Table 4.2, enzymes have a different sensitivity to PEF treatment. According to the literature, the factors that influence PEF enzymatic inactivation

TABLE 4.2. Summary of PEF-Enzyme Research.

Reference	PEF System	Results
Gilliland and Speck (1967)	• Electrohydraulic discharge • Aluminum core electrode • Electrode gap: 0.125 inch • 0.3 mM KH_2PO_4 solution or microorganism suspension • No temperature control	• *Lactic dehydrogenase:* At 31.5 kV/cm after 10 discharges, 100% reduction • *Trypsin:* At 31.5 kV/cm after 40 discharges, 50% reduction • *Proteinases of* B. subtilis *strain A:* At 31.5 kV/cm after 40 discharges, 60% reduction
Hamilton and Sale (1967)	• Rectangular direct current (d.c.) • Pulse length: 20 μsec • Frequency: 1 Hz • Carbon electrode • Polyethylene spacer with air contact • Temperature control by circulation of water at room temperature	• *NADH dehydrogenase, succinic dehydrogenase and hexokinase* in the extract of pulse treated *E. Coli* 8196: At 20–25 kV/cm, no significant inhibition • *Acetylcholinesterase* in pulse treated bovine erythrocytes, no inhibition • *Lipase and* α*-amylase* solution: At up to 30 kV/cm, no inhibition
Castro (1994)	• Gene electroporator (Kodak) • Exponential decay wave form • Frequency: 1/15 Hz • Pulse duration time: 0.74 msec or 0.4 msec • Disposable cuvette with air contact • Electrode gap: 0.1 cm • No temperature control	• *Alkaline phosphatase:* —In simulated milk ultrafiltrate: At 22 kV/cm after 70 pulses, 65% reduction —In raw milk and 2% milk: At 18.8 kV/cm after 70 pulses, 59% reduction
Vega et al. (1995a)	• Frequency: 0.1 Hz • Pulse duration time: 2 μsec • A continuous flow chamber • Two parallel stainless steel electrodes and a polysulfone spacer • Closed-loop system • Volume: 15 mL • Flow rate: 45 mL/min • Temperature control: cooling water circulation (15°C)	• *Plasmin* in SMUF (simulated milk ultrafiltrate): At 30 and 45 kV/cm after 50 pulses, 90% reduction
Vega et al. (1995b)	• Pilot plant pulser	• *Proteases from* Pseudomonas fluorescens: At 20 kV/cm, 10 pulses, 25% reduction At 20 kV/cm, 20 pulses,

TABLE 4.3. *Continued.*

		50% reduction At 35 kV/cm, 10 pulses, 60% reduction At 35 kV/cm, 20 pulses, 70% reduction

Grahl and
Märkl (1996)
	• High-voltage generator
	  with 5–15 kV d.c. voltage
	• Frequency: 1–22 Hz
	• Batch vessel BV-25
	  with air contact
	• Two plain parallel carbon
	  electrodes
	• Electrode gap: 0.5 cm
	• Electrode area: 50 cm²
	• Number of pulses: 1–20
	• No temperature control
		• Enzymes in raw milk:
		  At 21.5 kV/cm and high
		  energy input ($Q = 400$ kJ/L)
		  Lipase, 65% reduction
		  Peroxidase, 25% reduction
		  Alkaline phosphatase, <5%

Ho et al.
(1997)
	• High voltage pulse generator
	  (≤30 kV d.c.) with
	  instant charge reversal
	• Exponential decay wave form
	• Circular shape and batch
	  treatment chamber ≤148 mL
	• Two circular and parallel
	  stainless steel electrodes
	• Electrode gap: 0.3 cm
	• Horizontally positioned
	  chamber to avoid
	  electrical sparking
	• No cooling system
	• Pulse duration time: 2 μsec
	• Frequency: 0.5 Hz
	• Electrode distance: 0.3 cm
	• Number of pulses: 30
	• Enzyme in buffer solution
	  or deionized water
		• *Peroxidase:*
		  At 73.3. kV/cm, 30% reduction
		• *Alkaline phosphatase:*
		  At 83.3 kV/cm, 5% reduction
		• *α-Amylase:*
		  At 80 kV/cm, 85% reduction
		• *Lipase:*
		  At 88 kV/cm, 85% reduction
		• *Lysozyme:*
		  At 13.3 kV/cm, 60% reduction
		  At 50 kV/cm, 10% reduction
		
		• *Glucose oxidase:*
		  At 50 kV/cm, 75% reduction
		• *Polyphenol oxidase:*
		  At 50 kV/cm, 40% reduction
		• *Pepsin:*
		  At 40 kV/cm, 150% increase

Yeom et al.
(1999)
	• PEF bench-scale
	  processing unit
	• Stainless steel electrode
	• Electrode gap: 0.2 cm
	• Co-field flow, tubular chamber
	• Frequency: 1500 Hz
	• Pulse duration time: 4 μsec
	• Square wave and monopolar
	  pulse
	• Temperature control: 10°C
	• Number of pulses: up to 500
	• Enzyme in 1 mM EDTA solution
		• *Papain:*
		  —With activators
		   (L-Cys and DTT):
		   At 50 kV/cm,
		   50% reduction
		  —Without activators:
		   At 50 kV/cm, 90% reduction
		   after 24 hr storage at 4°C

61

can be summarized as follows:

(1) Electric parameters
  • electric field strength (Vega et al., 1995a, 1995b; Yeom et al., 1999)
  • total treatment time or number of pulses (Vega et al., 1995a, 1995b; Yeom et al., 1999)
  • pulse duration time and pulse width (Ho et al., 1997)
(2) Enzymatic structure
  • key components such as active site (Gilliland and Speck, 1967)
  • secondary structure (Ho et al., 1997; Yeom et al., 1999)
  • tertiary structure (Ho et al., 1997)
(3) Treatment temperature (Vega et al., 1995a)
(4) Suspension medium (Grahl and Märkl, 1996)

    PEF treatment does not cause any significant increase of temperature when a short pulse duration time is applied to a solution with low electric conductivity. Therefore, thermal inactivation of enzymes due to Joule heating can be excluded from the PEF enzymatic inactivation mechanism. When a cooling system is used, the nonthermal effect of PEF on enzyme activity is obvious (Vega et al., 1995a; Yeom et al., 1999). The pH of the enzyme solution is not changed after PEF treatment (Ho et al., 1997; Yeom et al., 1999). Therefore, reduction of enzyme activity by pH change is not the cause of PEF enzymatic inactivation. Conformational change of enzymes has been suggested as the PEF enzymatic inactivation mechanism by several researchers (Vega et al., 1995a; Ho et al., 1997; Yeom et al., 1999). To elucidate the PEF enzymatic inactivation mechanism, more information about the structure and function of enzymes under high voltage pulsed electric fields is necessary.

REFERENCES

Castro, A. J. 1994. Pulsed electric field modification of activity and denaturation of alkaline phosphatase. Ph.D. Dissertation. Washington State University.

Gilliland, S. E. and Speck, M. L. 1967. Mechanism of the bactericidal action produced by electrohydraulic shock. *Appl. Microbiol.* 15(5):1038.

Grahl, T. and Märkl, H. 1996. Killing of microorganisms by pulsed electric fields. *Appl. Microbiol. Biotechnol.* 45:148.

Hamilton, W. A. and Sale, A. J. 1967. Effects of high electric fields on microorganisms. *Biochimica et Biophysica Acta.* 148:789.

Ho, S. Y., Mittal, G. S., and Cross, J. D. 1997. Effects of high field electric pulses on the activity of selected enzymes. *J. of Food Eng.* 31:69.

Vega, H. M., Powers, J. R., Barbosa-Cánovas, G. V., and Swanson, B. G. 1995a. Plasmin inactivation with pulsed electric fields. *J. Food Sci.* 60(5):1143.

Vega, H. M., Powers, J. R., Barbosa-Cánovas, G. V., Swanson, B. G., and Luedecke, L. 1995b. Inactivation of a protease from *Pseudomonas fluorescens* M3/6 using high voltage pulsed electric fields. *IFT 1995. Annual Meeting*. Book of Abstracts, paper no. 89-3. p. 267.

Yeom, H. W., Zhang, Q. H., and Dunne, C. P. 1999. Inactivation of papain by pulsed electric fields in a continuous system. *Food Chem.* 67:53.

CHAPTER 5

Pulsed Electric Field Modification of Milk Alkaline Phosphatase Activity

A. J. CASTRO
B. G. SWANSON
G. V. BARBOSA-CÁNOVAS
Q. H. ZHANG

ABSTRACT

ALKALINE phosphate (ALP) from bovine milk was inactivated by pulses of high intensity electric fields (PEF). Alkaline phosphate is an indicator of the adequacy of thermal pasteurization of milk. Fluorometric analysis of ALP was used to assay the activity of ALP from raw whole milk, 2% milk, nonfat milk and modified simulated milk ultrafiltrate (MSMUF). Milk and MSMUF were treated with electric field intensities of 18.8 and 22.0 kV/cm, respectively. The activity of ALP was reduced by 65% in milk and MSMUF and 59% in raw milk and pasteurized, homogenized 2% milk. The temperature of milk treated with 70 pulses of 21.8 kV/cm increased from 22°C to 43.9°C. The temperature of MSMUF treated with 70 pulses of 22.3 kV/cm increased from 4°C to 8.4°C. The temperature increase did not contribute to the inactivation of ALP. Maximum velocity (V_{max}) and Michaelis constant (K_m) of ALP treated with PEF were determined. Seventy 0.46 msec pulses of 21.8 kV/cm reduced the V_{max} of ALP in milk from 51.4 to 42.9 mg of fluoroyellow/min/50 mL and increased the K_m of ALP in milk from 0.69 to 1.77. Seventy 0.78 msec pulses of 22.3 kV/cm increased the V_{max} of ALP in MSMUF from 13.3 to 24.4 mg of fluoroyellow/min/50 mL and increased the K_m of ALP in MSMUF from 0.12 to 0.54. Native ALP is resistant to trypsin proteolysis. ALP treated with seventy 0.78 pulses of 22.3 kV/cm was susceptible to partial proteolytic digestion by trypsin.

INTRODUCTION

The presence of alkaline phosphatase (ALP) in milk was described 62 years ago (Wilson and Hart, 1932). ALP in milk is sensitive to thermal denaturation, a finding employed to distinguish between raw milk and thermally pasteurized milk that contains no active ALP (Kay and Graham, 1933). The activity of ALP in pasteurized milk products is of public health significance, since the presence of active ALP indicates inadequate pasteurization or contamination of pasteurized milk with raw milk (Richardson et al., 1964; Murthy et al., 1993).

Milk alkaline phosphate (orthophosphoric-monoester phosphohydrolase, alkaline optimum, EC 3.13.1) is a dimeric metalloenzyme comprised of two very similar subunits of about 85,000 molecular weight (Linden and Alais, 1976). Milk alkaline phosphatase requires two metals for maximal activity: zinc is essential and magnesium is stimulatory (Linden et al., 1977). Alkaline phosphatase contains 4.0 g-atoms of zinc per mole of protein (Linden and Alais, 1976). The mechanism of action of ALP is of the flip-flop type (Lazdunski et al., 1971). Mg^{2+} induces the asymmetry the flip-flop mechanism requires (Linden et al., 1977). Mg^{2+} cannot replace Zn^{2+} at the zinc sites to give an active ALP, and Zn^{2+} becomes an inhibitor when fixed on the magnesium site (Linden and Alais, 1978). Mg^{2+}, Mn^{2+}, and Ca^{2+} ions stimulate ALP activity in milk (Sharma and Ganguli, 1970). Inorganic phosphate is a substrate, a product and an inhibitor of ALP activity (Morton, 1955; Linden, 1979; Wyckoff et al., 1983).

In milk pasteurized at high temperatures for short periods of time (161°F/15 s), ALP is inactivated, but often regenerates during storage (Richardson et al., 1964; Wright and Tramer, 1954; Brown and Elliker, 1942). ALP regeneration was first observed in pasteurized cream (Linden, 1979). Wright and Tramer (Sharma and Ganguli, 1970; Morton, 1955; Wyckoff et al., 1983; Wright and Tramer, 1954, 1956) reported that the aging of milk prior to pasteurization, the absence of air, pasteurization at high temperatures for short periods of time, and the presence of magnesium and calcium enhanced regeneration. Regenerated alkaline phosphatase is similar to active alkaline phosphatase as judged by thermal stability, pH optimum and rate of phosphate hydrolysis at selected pHs (Wright and Tramer, 1953a, 1953b). Peereboom (Peereboom, 1966; 1968) reviewed the concept of ALP regeneration and concluded that raw and regenerated ALP are not identical. Native alkaline phosphatase from raw cream exhibits an electrophoresis pattern of at least three isoenzymes with alkaline phosphatase activity. The isoenzyme electrophoresis pattern of raw milk separates α, β, and γ isoenzymes. The electrophoresis pattern of pasteurized milk provides only the β isoenzyme. During pasteurization of milk, a part of the alkaline phosphatase isoenzyme is not only temporarily inactivated, but also structurally changed. After regeneration of ALP in pasteurized cream only regenerated β isoenzymes were observed in the electrophoresis pattern (Peereboom, 1968).

The first of the ALP reactions results in formation of the alkaline phosphatase substrate (Michaelis) complex (McComb et al., 1979). Evidence presented by Daemen and Riordan (1974) suggests that one arginine molecule is directly involved in the substrate binding process at each of the two binding sites of ALP. Chemical modification of arginine by two specific arginine reagents, 2,3-butanedione and phenyl glyoxal produces inactivation of alkaline phosphatase (Daemen and Riordan, 1974). Magnesium and zinc also participate in the ALP binding process to inorganic phosphate, the substrate of alkaline phosphatase (Peereboom, 1968). The formation and stability of the ALP Michaelis complex are elucidated indirectly by the measurement of the Michaelis constant (K_m), the phosphate concentration at which the rate of the ALP catalyzed reaction proceeds at one-half the maximum rate (Stryer, 1988). After the formation of the ALP-substrate complex, a phosphoprotein, phosphoryl enzyme with a phosphate group covalently bound to the ALP is formed, while the alcohol (ROH) group is released into the medium [Equation (1)] (McComb et al., 1979). The release of a quantity of ROH is assayed during ALP reactions to determine the reaction rate (Rocco, 1990a).

$$
\begin{array}{ccccc}
\overset{\displaystyle O^-}{\underset{\displaystyle O^-}{|}}\\
\text{EH} + \text{ROP} = O & \Longleftrightarrow & \text{EH} \cdot \text{P} - \text{OR} & \Longleftrightarrow & \text{E} - \text{P} - O^- + \text{ROH} \quad (1)
\end{array}
$$

Electric fields affect molecular structure in three ways: (1) the movement of free electrons, ions, and other charged particles is stimulated; (2) bound charges, electrons in atoms, and atoms in molecules are displaced to form polar particles; and (3) protein and water molecules with a constant dipole moment are oriented (Presman, 1970; De Maeyer, 1969). The results of the electric field action on molecules and electrical-chemical coupling can be summarized as follows: (a) polar structures tend to orient in the direction of the applied electric field; (b) protein conformations and molecules with large dipole moments increase in concentration at the expense of molecules with small dipole moments; and (c) the presence of electric fields increases the dissociation of weak acids and bases, and promotes the separation of ion pairs into corresponding dissociated ions or ionic groups (Neumann, 1981). The dissociation of weak acids and bases, by electric fields is called the second Wien effect (Schelly and Astumian, 1984). When molecules carrying permanent or induced dipole moments undergo chemical transformations, three types of electrical chemical coupling are considered: permanent dipole equilibria, induced dipole equilibria, and ionic dissociation (De Maeyer, 1969; Neumann, 1981).

The alkaline phosphatase protein molecule is approximately 45% α-helix, 13% β-sheet and 16% β-turn (Coleman and Gettins, 1983). An amino acid residue, alternatively called a peptide unit, is an electric dipole of 3.46 D (Deby) (Hol et al., 1978). In an α-helix, dipoles are aligned in the same direction; i.e., the positive pole of the dipoles points to the N-terminus of the peptide, and the negative pole of the dipoles points to the C-terminus of the peptide (Hol et al., 1978). Thus, an α-helix composed of n amino acids contains a dipole of $3.46 \times n \times$ D. β-Sheets, on the other hand, exhibit little net dipole moment (Hol et al., 1981). Water dipoles may also interact with dipoles, created by an external electric field, from the peptide unit (Presman, 1970). Negatively charged phosphate may be electrostatically attracted to active sites by partial positive charges located at the N-terminals of the helix due to dipole moments of the helices (Hol et al., 1978).

Pulses of high intensity electric fields (PEF) are a potentially important cold pasteurization/stabilization food preservation technique that may replace, or partially substitute for, thermal processing techniques (Castro et al., 1993). Inactivation of microbes in liquids is effectively performed by PEF. Bacteria, yeast, and mold cells in milk, yogurt, orange juice and fluid eggs are reduced five log cycles by PEF (Castro et al., 1993). Lipase and vitamin C in PEF treated milk are inactivated, but alkaline phosphatase in milk is not inactivated by PEF (Grahl et al., 1992). Gilliland and Speck (1967), using an electrohydraulic treatment, reported that enzymes containing or requiring free sulfhydryl groups such as reduced glutathione are most susceptible to electric fields. Hamilton and Sale (1967) reported no inhibition of NADH dehydrogenase, succinic dehydrogenase, or hexokinase in high intensity pulsed electric fields.

The objective of this research was to study the effect of high intensity pulsed electric fields on the activity and structure of alkaline phosphatase in milk and modified simulated milk ultrafiltrate.

MATERIAL AND METHODS

PULSED ELECTRIC FIELDS

Alkaline phosphatase type XXI from bovine milk, 3.5 units/mg, was purchased from Sigma Chemical Co. (St. Louis, MO) and dissolved at 2 mg/mL in raw milk, 2% milk, nonfat milk, or modified simulated milk ultrafiltrate (MSMUF) (Jennes and Koops, 1962).

MSMUF is a formulated system of milk constituents used in nonthermal processing research to study the properties of milk proteins and potential changes during nonthermal processing. Simulated milk ultrafiltrate (SMUF) is a salt solution with a selected composition encountered in milk ultrafiltrates (Jennes and Koops, 1962). Lactose (5%) was added to depress the electrical conductivity of the simulated milk ultrafiltrate (MSMUF) to 500 to 600 μmhos/cm. MSMUF was adjusted to the optimum pH of 8.0 for ALP activity. MSMUF was passed through filters of 0.22 μm pore width (Falcon, Becton Dickinson Labwere, Lincoln Park, NJ) to remove insoluble and denatured materials.

The alkaline phosphatase dissolved in MSMUF was transferred to a sterile and disposable cuvette with a 0.1 cm interelectrode distance (BioRad Lab., Hercules, CA). High intensity electric field pulses were applied with a GeneZapper (Kodak, New Haven, CT) capable of generating electric field intensities up to 2500 V when using a 7 μfarads capacitor. A 5 Ω resistor was placed in series with the cuvette to increase the electric field intensity. An exponential decay waveform pulse was produced by discharge of the capacitor. The electric field intensity declines over time as a function of the resistance in the circuit and the size of the capacitor. The exponential decay curve is described by:

$$E_{(t)} = E_0 e^{-t\iota} \qquad (2)$$

Where $E_{(t)}$ is the electric field intensity ($\mathrm{kVcm^{-1}}$) at any time t (s). E_0 is the initial field intensity, and ι is the resistance capacitance (RC) time constant (s). The time constant depends on the total resistance (R in Ω) and capacitance (C in F) of the system as follows:

$$\iota = RC \qquad (3)$$

Therefore, ι describes the shape of the decay waveform and is the time required for the electric field intensity to decline to 37% of the initial value. Because the resistance of the solution affects the time constant, the composition of the solution and the geometry of the cuvette influence the shape of the discharge pulses. The resistance of the solution is inversely proportional to the ionic strength, temperature, and cross-sectional area of the solution-electrode interface. The resistance is directly proportional to the distance between the electrodes.

A maximum of 70 exponential decay pulses of 0.40 milliseconds for milk and 0.74 milliseconds for MSMUF were applied. Time between pulses was 15 s. The electric field intensity of 13.2 to 18.8 $\mathrm{kVcm^{-1}}$ was selected for MSMUF. A parallel resistance of 100 ohms and a capacitance of 7 μfarads were selected for each PEF treatment.

FLUOROMETRIC DETERMINATION OF ALKALINE PHOSPHATE ACTIVITY

Alkaline phosphatase (ALP) activity was determined with the fluorophos ALP assay (Rocco, 1990a, 1990b), a newly approved Association of Official

Analytical Chemistry fluorometric assay (AOAC 991.24) used by Eckner (Eckner, 1992) to study the relationship between ALP inactivation and bacterial pathogen inactivation in raw milk. Eckner (1992) demonstrated the relationships among ALP, *Listeria monocytogenes,* and *Salmonella senftenberg* inactivation by thermal pasteurization. The fluorophos ALP assay is more accurate than the standard Scharer rapid phosphatase test (Scharer, 1938), and produces results within three minutes (Quicker results, 1994). In the fluorophos ALP assay, active ALP in the tested solution hydrolyzes the fluorophos substrate to highly fluorescent fluoroyellow.

Fluoroyellow formation was monitored continuously over 3 min, and the ALP activity in milliunits per liter was calculated from the recorded increase in fluorescence of the solution per minute. A spectrophotofluorometer (J4-8961-Aminco-Bowman, Silver Spring, MA), was used to monitor the formation of the fluoroyellow. The cuvette was maintained at a constant temperature of 22°C. An excitation wavelength of 439 nm and emission wavelength of 560 nm were selected. The cuvettes were quartz cell type 3H (NSG Precision Cells, Inc., Farmingdale, NY). Fluorophos was purchased from Advanced Instruments Inc. (Norwood, MA).

The analytical procedure was as follows. After the electrical treatment, 75 microliters of well-mixed ALP in raw milk, 2% milk, nonfat milk, or MSMUF were added to 2 mL of fluorophos in a 12×75 mm test tube. The solutions were immediately mixed by gentle inversion, poured into the quartz cuvette and the cuvette placed into the fluorometer. After 1 min of temperature equilibration, the rate of increase in fluorescence was recorded (fluoroyellow/min) over the next 2 min.

DETERMINATION OF K_m VALUES AND V_{max} OF ALKALINE PHOSPHATASE

The Michaelis constant (K_m) and maximum velocity (V_{max}) of ALP in milk and MSMUF treated or not treated (control) with PEF were determined. Fluorophos substrate concentrations ranging from 200 to 2000 μL for milk (Rocco, 1990b) and MSMUF were selected. K_m values of ALP were calculated from a Lineweaver-Burk double reciprocal plot (Lineweaver and Burk, 1934) based on the rearrangement of the Henri-Michaelis-Menten equation into a linear ($y = mx + b$) form. On a plot of $1/[S]$ the slope is equivalent to K_m/V_{max} and the intercept on the y axis ($1/v$) is equivalent to $1/V_{max}$. When $1/v = 0$, $1/[S]$ is equivalent to $-1/K_m$ (Segel, 1975). Computation of an equation for the best fitting line was obtained from the statistical analysis (SAS, 1987). The correlation coefficient, r, was determined by ANOVA using the SAS system ($\alpha = 0.05$) (SAS, 1987).

DIGESTION OF ALP BY TRYPSIN

One millimeter of ALP in MSMUF was treated with seventy 0.78 msec pulses of 22.3 kV/cm, and mixed with 2 mL of 0.1 mg/mL trypsin solution (Schlesinger, 1965) in 0.1 M phosphate buffer, pH 7.5. The solution was incubated at 30°C throughout the assay. Untreated and treated ALP in MSMUF were digested for 20 min. The optical density (*OD*) of untreated or PEF treated ALP in MSMUF at 278 nm (Okunuki, 1961) was assayed in a diode array spectrophotometer (Hewlett Packard, 8452A, Waldbronn, FRG).

The ultraviolet absorption spectrum of untreated ALP in MSMUF exhibited a maximum at 278 nm. The maximum absorbance of ALP in MSMUF after PEF treatment was assayed after 20 min of incubation with trypsin. An increase in the susceptibility of ALP to trypsin hydrolysis was used to estimate the ratio of inactivation (*RI*) of ALP (Okunuki, 1961). The ratio of the inactivation was calculated from the following equation:

$$RI\% = \frac{A' - A''}{A} \times 100 \qquad (4)$$

where A' is the initial absorbance of treated ALP, A'' is the final absorbance of treated ALP, and A is the absorbance of native ALP under equivalent conditions.

THERMAL METHODOLOGY

The change in temperature of 100 μL aliquot of a milk mixture prepared from 1 mL of raw milk containing 100% active ALP in 100 mL UTH pasteurized 2% milk treated by seventy 0.45 msec pulses of 13.2 kV/cm PEF was assayed at an initial temperature of 22°C. The maximum temperature reached by the milk in the cuvette was determined by inserting a thermocouple (Omega T-TT-40, Samford, CT) in the electrode gap immediately after the seventieth pulse.

Five milliliter glass tubes containing 1 mL of milk mixture containing ALP were incubated at 43.9°C for a holding time of 17.5 min, equivalent to the time needed to apply 70 pulses. After 17.5 min incubation, the ALP activity of the milk was determined by the fluorophos ALP assay (Rocco, 1990a). Aliquots of milk were removed at intervals equivalent in time to 10 pulses at 43.9°C. The first aliquot of milk was removed after 2.5 min and the last aliquot removed after 17.5 min, equivalent in time to 70 pulses.

A second experiment was conducted in a cold room at 4°C. A thermocouple (Omega T-TT-40, Samford, CT) was connected to the ground electrode of the Gene-Zapper and the change of MSMUF solution temperature in the curvette was assayed immediately after each pulse. Fifteen, 30 and 60 s intervals between pulses were selected for the experiment. The temperature of 100 μL of 1 mg/mL ALP in MSMUF in the cuvette was stabilized to 4°C for 5 min, and used as the initial temperature. A current of cold air was circulated into the reaction module of the GeneZapper to maintain 4°C around the cuvette. After seventy 0.78 msec pulses of 23.0 kV/cm, the cuvette with 100 μL of 1 mg/mL treated ALP in MSMUF was placed in an ice bucket and the ALP activity immediately assayed with a spectrophotofluorometer.

Heat induced by one 0.78 msec pulse of 23.0 kV/cm treated bovine ALP dissolved in MSMUF was calculated using the voltage (V) and current (I) collected with digitizing oscilloscope (Hewlett Packard 54520A, 500 MS a/s 500 mHz, Colorado Springs, CO). The values calculated from voltage and current were time constant, resistance of the MSMUF (R) in ohms, temperature (T) in °C, and energy (E) in joules:

$$R = V/I \tag{5}$$

$$T = T_0 + E_m/\rho c_p v \tag{6}$$

where ρ is the density (1 g/cm^3), c_p is the specific heat (4.2 J/g°C) and v is the volume (0.1 cm^3). E_m is the energy calculated from:

$$E_m = \frac{1}{2}(CV)^2 \tag{7}$$

where C is the capacitance and V is the charging voltage.

RESULTS AND DISCUSSION

ALKALINE PHOSPHATASE ACTIVITY

Inactivation of alkaline phosphatase by pulsed electric fields depends on the PEF treatment electric field intensity, the amount of fat in the milk, and the concentration of ALP. As the field intensity applied to milk or MSMUF increases, the activity of ALP decreases. Seventy 0.45 msec pulses of an electric intensity of 14.8 kV/cm applied to 2% milk reduces the ALP activity by 43%. Seventy 0.40 msec pulses of a field intensity of 18.8 kV/cm applied to 2% milk reduces the ALP activity by 59% (Figure 5.1). Seventy 0.74 msec pulses of a field intensity of 22.0 kV/cm applied to 2 mg/mL ALP dissolved in MSMUF reduces the ALP activity by 65%.

The activities of ALP dissolved in UTH pasteurized 2% and 4% milk were reduced 59% by seventy 0.40 msec pulses of 18.8 kV/cm and the activity of ALP dissolved in nonfat milk was reduced by 65% (Figure 5.2).

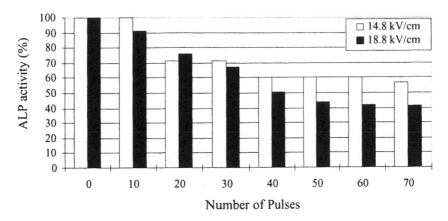

Figure 5.1 PEF inactivation of ALP diluted in UHT pasteurized 2% milk.

HEAT AND PEF

Milk treated with an electric field intensity of 13.2 kV/cm at 22°C increased in temperature to 43.9°C after the seventieth pulse. The activity of natural ALP milk obtained by mixing 1 mL of raw milk in 100 mL of 2% milk was reduced by 96% after the milk was treated with pulsed electric fields, and by 30% after it was heated at 43.9°C for 17.5 min (Figure 5.3).

ALP (1 mg/mL) dissolved in MSMUF was treated with seventy 0.78 msec pulses of 23.0 kV/cm at 4.0°C. Activity of the ALP decreased by 32.6% when treated at intervals of 25 s between pulses, whereas at intervals of 30 and 60 s, the activity decreased by 37.5% and 43.3%. When the interval between pulses

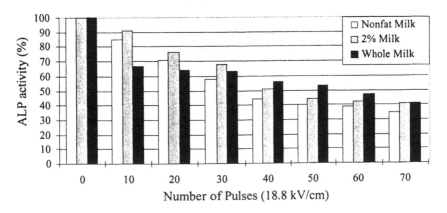

Figure 5.2 Inactivation of ALP diluted in UHT pasteurized nonfat, 2%, and whole milk.

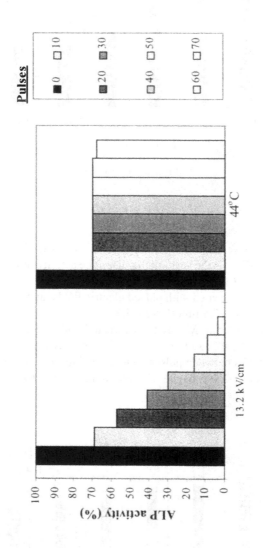

Figure 5.3 Inactivation of alkaline phosphatase by PEF or 44°C for 17.5 min.

was 15 s, the ALP [(1 mg/mL) in MSMUF solution] temperature increased from an initial temperature of 4.0°C to 8.4°C after the seventieth pulse. When the intervals were 30 and 60 s, the solution temperature increased from an initial temperature of 4.0°C to 6.3°C and 5.3°C, respectively.

The temperature of the solution of ALP in MSMUF increased from an initial temperature of 4.0°C to 16.7°C after one 0.78 msec pulse of 23 kV/cm. The calculated electrical energy was 5.3 Joules per pulse and the current was 8.7 amperes. The resistance of the solution of ALP decreased from 16 to 4 ohms at the end of the pulse; the final temperature of ALP in MSMUF was 16°C, 12°C greater than the initial temperature. The initial resistance was 262.1 ohms, and at the final resistance of ALP (1 mg/mL) dissolved in MSMUF was 150.1 ohms. The application of PEF to milk or to MSMUF increases the temperature. The increase of the temperature was due, mainly, to Joule heating. Joule heating is electrical energy dissipated as heat in the medium acting as a resistor.

The generator heat is absorbed by the electrodes of the curvette. The electrical resistance at 4°C was ~280 ohms (Figure 5.4). The electric field intensity at 4°C was 23 kV/cm, 0.7 kV/cm greater than when the MSMUF was treated at 22°C.

Inactivation of the ALP is directly proportional to the electric field intensity. As a result, when the ALP (1 mg/mL) in MSMUF solution was treated with PEF at 4°C, the inactivation of ALP was greater than when the PEF treatment was applied at 22°C. The inactivation of ALP at 4°C and 22°C demonstrated that the temperature of the MSMUF during the PEF treatment increased the electric field intensity, and that heat induced by the electric fields did not participate in the inactivation of the ALP during high voltage electric field treatment.

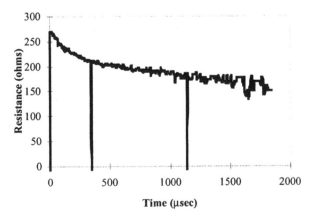

Figure 5.4 Electrical resistance of ALP in MSMUF changes as a function of time.

TABLE 5.1. Pulsed Electric Fields and the Hydrolysis of Fluorophos by Alkaline Phosphatase at 22°C.

	Pulses	$1/V_{max}$	$1/K_m$	V_{max}	K_m	Slope
Milk	0	1.95E − 02	−1.45E + 00	5.14E + 01	6.88E − 01	1.34E − 02
	35	3.01E − 02	−1.22E + 00	3.32E + 01	8.22E − 01	2.47E − 02
	70	2.33E − 02	−5.67E − 01	4.29E + 01	1.77E + 00	4.11E − 02
MSMUF	0	7.50E − 03	−8.65E + 00	1.33E + 02	1.16E − 01	8.70E − 04
	35	6.10E − 03	−4.19E + 00	1.63E + 02	2.39E − 01	1.47E − 03
	70	4.10E − 03	−1.86E + 00	2.44E + 02	5.37E − 01	2.20E − 03

V_{max} AND K_m

Alkaline phosphatase activity of milk and MSMUF treated with 35 and 70 pulses of high intensity electric fields was determined at 22°C. When $1/V$ was plotted against 1/fluorophos concentration according to the method of Lineweaver-Burk (Lineweaver and Burk, 1934), straight lines were obtained. The slope, K_m/V_{max}, of the lines is presented in Table 5.1. The straight lines describe the response of the reciprocal velocity to the reciprocal independent variable substrate concentration. Fluorophos is the substrate for ALP.

The Lineweaver-Burk plots presented lines intersecting to the left or right of the $1/V$ axis, depending upon whether the ALP was from whole milk or MSMUF (Figures 5.5 and 5.6).

The maximum rate of the fluoroyellow producing reaction (V_{max}) for ALP in milk decreased when the milk was treated with PEF, and the binding of ALP to fluorophos (K_m) increased with an increasing number of 0.46 msec pulses

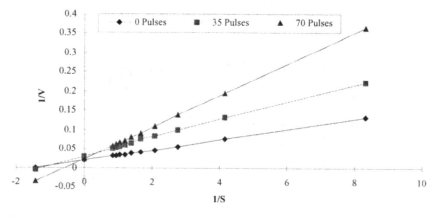

Figure 5.5 Lineweaver-Burk plot of bovine milk ALP treated with 0, 35, and 70 pulses of 0.46 msec pulse duration.

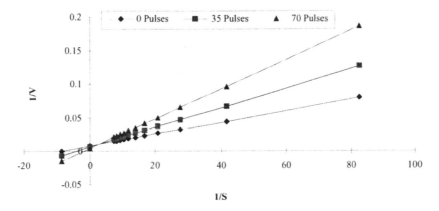

Figure 5.6 Lineweaver-Burk plot of ALP diluted in MSMUF treated with PEF.

of 21.8 kV/cm (Figures 5.5 and 5.7). A decrease in V_{max} and an increase in K_m characterizes type I enzyme inhibition. $V_{max}'^1 < V_{max}^0$ and $K_m^{1'} > K_m^0$ are designated as two parameter matched inhibition (Krupyanko, 1988), traditionally referred to as mixed inhibition. Alkaline phosphatase treated with pulsed electric fields (ALP') displays two parameter matched inhibition when ALP' exhibits less affinity for fluorophos than untreated ALP, and the ALP'-fluorophos complex produces smaller amounts of fluoroyellow.

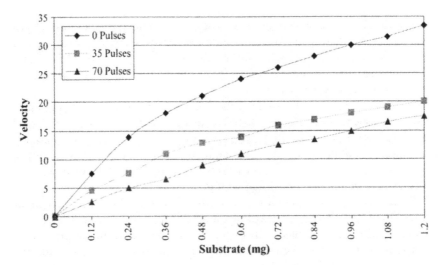

Figure 5.7 Plot of the initial velocity of fluoroyellow (FY) producing reaction of milk ALP treated with 0.46 msec pulses of PEF vs. the fluorophos concentration.

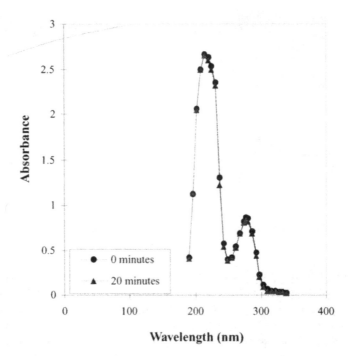

Figure 5.8 Ultraviolet absorption of native ALP after incubation with trypsin for 20 min.

DIGESTION OF ALP BY TRYPSIN

Native ALP is resistant to trypsin proteolysis, maintaining a constant absorbance value during 20 min of incubation. ALP is susceptible to trypsin proteolysis after PEF treatment with seventy 0.78 msec pulsed of 22.3 kV/cm, the absorbance at 278 nm decreasing from 1.62 to 1.24 (23.5%). The ratio of ALP inactivation (RI) was 44%.[1]

Decrease in biological activity is used to determine enzyme denaturation, since previous research suggests that enzymatic activity is lost when enzymes are denatured (Kauzmann, 1959). Enzymes may be inactivated in the presence of a specific inhibitor without being denatured. An increase in the susceptibility of the alkaline phosphatase to trypsin proteolysis indicated degradation of the secondary structure of ALP treated with PEF.

Schlesinger (1965) reported a shift in the ultraviolet absorption maximum from 276 to 279 nm of *E. coli* ALP upon acidification and trypsin proteolysis. The red shift was attributed to perturbation of tyrosine resulting after tyrosine removal from an apolar environment. The difference in the maximum absorbance of ALP at 220 nm before and after PEF treatment (Figures 5.8 and 5.9)

---

[1] K'_m and $V'_{\max}$ are the effective Michaelis constant and maximum reaction rate determined in the presence of PEF; K^0_m and $V^0_{\max}$ are the same parameters of the control, untreated.

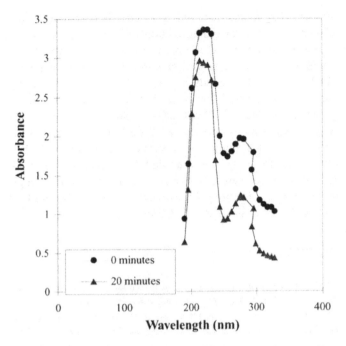

Figure 5.9 Ultraviolet absorption of alkaline phosphatase treated with seventy 0.78 msec pulses of 22.3 kV/cm after incubation at 30°C for 20 min with trypsin.

may be attributed to peptide unfolding (Krupyanko, 1990). Kauzmann (1959) ascribed the spectral red shifts observed in protein denaturation to changes in the electric charges near tyrosine residues of the protein. An external electric field influences the conformational state of a protein through charge, dipole, or induced dipole chemical reactions (Tsong and Astunian, 1986).

The maximum inactivation of ALP in milk or MSMUF was 65%, obtained respectively with seventy 0.40 msec pulses and a field intensity of 18.8 kV/cm, or seventy 0.70 msec pulses and a field intensity of 22.0 kV/cm. Grahl et al. (1992) reported no inactivation of ALP in milk after twenty 0.39 μsec pulses and a field intensity of 26.0 kV/cm. The longer pulse duration and increased field intensity explain the inactivation of ALP. Inactivation of alkaline phosphatase by pulsed electric fields is directly related to the number of pulses and the intensity of the electric field.

CONCLUSIONS

Pulses of high voltage electric fields inactivate alkaline phosphatase. The maximum inactivation of ALP was 65%. The inactivation of ALP is directly related to the concentration of the ALP and to the intensity of the electric field.

Milk and MSMUF treated with pulsed electric fields reached a temperature of 43.9°C when the initial temperature of the milk and MSMUF was 22°C, and reached a temperature of 8.4°C when the initial temperature of the milk and MSMUF was 4°C. The increase in temperatures did not affect the inactivation of ALP. PEF changed ALP, producing an ALP that either reacted producing smaller amounts of fluoroyellow, or did not react with substrate fluorophos, affecting the reaction rate even when the fluorophos was at a large concentration. The increase in K_m demonstrates that ALP treated with pulses of 22.3 kV/cm of PEF requires a larger amount of substrate than untreated ALP to achieve maximal catalytic efficiency, indicating that the affinity of treated ALP for the fluorophos substrate decreased. Native ALP is resistant to trypsin proteolysis. ALP was susceptible to the proteolytic digestion of trypsin after PEF treatment with seventy 0.78 msec pulses of 22.3 kV/cm.

REFERENCES

Brown, W. H. and Elliker, P. R. 1942. The effect of flash pasteurization and subsequent treatment of the phosphatase value of cream. *Journal of Bacteriology,* 43, 118–119.

Castro, A. J. Barbosa-Cánovas, G. V., and Swanson, B. G. 1993. Microbial inactivation of foods by pulsed electric fields. *Journal of Food Processing and Preservation,* 17, 47–73.

Coleman, J. E. and Gettins, P. 1983. Alkaline phosphatase, solution structure, and mechanism. *Advances in Enzymology,* 45, 382–452.

Daemen, F. J. and Riordan, J. F. 1974. Essential arginyl residues in *E. coli* alkaline phosphatase. *Biochemistry,* 13, 2865–2871.

De Maeyer, L. C. M. 1969. Electric field methods. *Methods in Enzymology,* 14. Academic Press, NY. pp. 80–118.

Eckner, F. K. 1992. Fluorimetric analysis of alkaline phosphatase inactivation correlated to *Salmonella* and *Listeria* inactivation. *J. Food Protect.,* 55, 960–963.

Gilliland, S. E. and Speck, M. L. 1967. Mechanisms of the bactericidal action produced by electrohydraulic shock. *Applied Microbiology,* 15, 1038–1044.

Grahl, Th., Sitzmann, W., and Markl, H. 1992. Killing of microorganisms in fluid media by high-voltage pulses. *Dechema Biotechnology Conferences,* 5B, 675–678.

Hamilton, W. A. and Sale, A. J. H. 1967. Effects of high electric fields on microorganisms I. Mechanism of action of the lethal effect. *Biochimica and Biophysica Acta,* 148, 789–800.

Hol, W. G. J., Halie, L. M., and Sander, C. 1981. Dipoles of the α-helix and β sheet: Their role in protein folding. *Nature,* 294, 532–536.

Hol, W. G. J., van Duijen, P. T., and Berendsen, H. J. C. 1978. The α-helix dipole and the properties of proteins. *Nature,* 273, 3–446.

Jennes, R. and Koops, J. 1962. Preparation and properties of salt solution which simulates milk ultrafiltrate. *Netherland Milk and Dairy Journal,* 16, 153–164.

Kauzmann, W. 1959. Some factors in the interpretation of protein denaturation. Advances in protein. *Chemistry,* 14, 1–63.

Kay, H. D. and Graham, W. H. 1933. Phosphorus compounds of milk. *Journal of Dairy Science,* 5, 65–74.

Krupyanko, V. I. 1988. A vector method of representing individual types of enzymic reactions in $K'_m V'$ coordinates. *Collection of the Czechoslovak Chemists Community*, 53, 161–172.

Krupyanko, V. I. 1990. On linear concentration dependences of $K'_m V'$ and $1/v_1$ enzyme kinetic parameters. *Collection of the Czechoslovak Chemists Community*, 55, 1351–1365.

Lazdunski, M., Peticlerc, C., Chappelet, D., and Lazdunski, C. 1971. Flip-Flop mechanism in enzymology. A model: The alkaline phosphatase of *E. coli*. *European Journal of Biochemistry*, 20, 124–139.

Linden, G. 1979. Biochemical study of some aspects of milk alkaline phosphatase reactivation. *Milchwissenschaft*, 34, 329–332.

Linden, G. and Alais, C. 1976. Phosphatase alcaline du lait de vache. II. Structure sous-unitaire, nature metalloproteique et parametres cinetiques. *Biochimica et Biophysica Acta*, 129, 205–213.

Linden, G. and Alais, C. 1978. Alkaline phosphatase in human, cow and sheep milks: Molecular and catalytic properties and metal ion action. *Ann. Biol. Anima. Bioch. Biophys.*, 18, 749–758.

Linden, G., Chappelet-Tordo, D., and Lazdunski, M. 1977. Milk alkaline phosphatase stimulation by Mg^{2+} and properties of the Mg^{2+} site. *Biochimica et Biophysica Acta*, 483, 100–106.

Lineweaver, H. and Burk, D. 1934. The determination of enzyme dissociation constant. *Journal of the American Chemical Society*, 56, 658–666.

McComb, R. B., Bowers, Jr., G. N., and Posen, S. 1979. *Alkaline Phosphatase*. Plenum Press, NY. pp. 229–288.

Morton, R. K. 1955. Some properties of alkaline phosphatase of cow's milk and calf intestinal mucous. *Biochemical Journal*, 60, 573–582.

Murthy, G. K., Klein, D. H., Richardson, T., and Rocco, R. M. 1993. Alkaline phosphatase methods. In: *Standard Methods for the Examination of Diary Products*, 16th ed. Marchal, R. T. Ed., Am. Public. Health Assoc., Washington, DC. pp. 413–431.

Neumann, E. 1981. Principles of electric fields effects in chemical and biological systems. *Topics of Bioelectrochemistry and Bioenergetics*, 4, 114–160.

Okunuki, K. 1961. Denaturation and inactivation of enzyme proteins. *Advances in Enzymology*, 23, 30–82.

Peereboom, J. W. C. 1966. Studies on alkaline phosphatase I. Non-identity of raw and reactivated alkaline phosphatase from cream. *Netherland Milk and Dairy Journal*, 20, 113–122.

Peereboom, J. W. C. 1968. Studies on alkaline phosphatase II. Occurrence of various phosphatase isoenzymes in dairy products. *Netherland Milk and Dairy Journal*, 22, 137–152.

Presman, A. S. 1970. *Electromagnetic Fields and Life*. Plenum, NY. pp. 25–64.

Quicker results. 1994. *Dairy Field*, 177, 64.

Richardson, L. A., McFarren, E. F., and Campbell, J. E. 1964. Phosphatase reactivation. *Journal of Dairy Science*, 47, 205–210.

Rocco, R. M. 1990a. Fluorimetric determination of alkaline phosphatase in fluid dairy products: Collaborative study. *Journal of the Association of Official Analytical Chemists*, 73, 842–849.

Rocco, R. M. 1990b. Fluorimetric analysis of alkaline phosphatase in fluid dairy products. *Journal of Food Protection*, 53, 588–591.

SAS Institute Inc. 1987. *SAS/Stat. Guide for Personal Computer*. Version 6 edition. Cary, NC.

Scharer, H. 1938. A rapid phosphomonoesterase test for control of dairy pasteurization. *Journal of Dairy Science*, 21, 21–34.

Schelly, Z. A. and Astumian, R. D. 1984. A theory for the apparent "negative second Wien effect" observed in electric-field-jump studies of suspensions. *The Journal of Physical Chemistry*, 88, 1152–1156.

Schlesinger, M. J. 1965. The reversible dissociation of the alkaline phosphatase of *E. coli*. II. Properties of the subunit. *Journal Biological Chemistry*, 240, 4293–4298.

Segel, H. I. 1975. *Enzyme Kinetics*. John Wiley and Sons, NY. pp. 44–48.

Sharma, R. S. and Ganguli, N. C. 1970. Effect of milk constituents on purified alkaline phosphatase from buffalo and cow milk. *Indian Journal of Biochemistry*, 7, 285–286.

Stryer, L. 1988. *Biochemistry*. 3rd Edition. W. H. Freeman, NY.

Tsong, T. Y. and Astunian, R. D. 1986. Absorption and conversion of electric field energy by membrance bound ATPases. *Bioelectricity and Bioenergetics*, 15, 457–474.

Wilson, L. T. and Hart, E. B. 1932. The chemistry of the blood of dairy cows before and after pasteurization and its relation to milk fever. *Journal of Dairy Science*, 15, 116–121.

Wright, R. C. and Tramer, J. 1953a. Reactivation of milk phosphatase following heat treatment. I. *Journal of Dairy Research*, 20, 177–188.

Wright, R. C. and Tramer, J. 1953b. Reactivation of milk phosphatase following heat treatment. II. *Journal of Dairy Research*, 20, 258–272.

Wright, R. C. and Tramer, J. 1954. Reactivation of milk phosphatase following heat treatment. III. *Journal of Dairy Research*, 21, 37–49.

Wright, R. C. and Tramer, J. 1956. Reactivation of milk phosphatase following heat treatment. IV. *Journal of Dairy Research*, 23, 248–256.

Wyckoff, H. W., Handschumacher, M., Murthy, H. M., and Sowadski, J. M. 1983. The three dimensional structure of alkaline phosphatase from *E. coli*. *Advances in Enzymology*, 45, 453–480.

CHAPTER 6

Pulsed Electric Field Denaturation
of Bovine Alkaline Phosphatase

A. J. CASTRO
B. G. SWANSON
G. V. BARBOSA-CÁNOVAS
A. K. DUNKER

ABSTRACT

PULSES of high intensity electric fields (PEF), ~20 kV/cm, are potentially an important nonthermal pasteurization/sterilization food preservation technique. Foods with a relatively low conductance, such as milk, yogurt, and liquid eggs, appear amenable to PEF treatment. The objective of the present research is to investigate the structural, physical, and chemical changes of alkaline phosphatase treated by PEF.

Alkaline phosphatase (ALP) from bovine milk was subjected to an electric field of 22.0 kV/cm with pulses of 0.7–0.8 msec. PEF treated ALP was assayed with 90° light scattering, intrinsic tryptophan fluorescence, and extrinsic fluorescence arising from two hydrophobic probes, 1-anilinonaphthalene-8-sulfonate (ANS) and cis-parinaric acid (CPA).

Polyacrylamide gel electrophoresis (PAGE) of ALP treated with seventy 0.70 msec pulses of 20.8 kV/cm revealed a single protein band with mobility identical to the untreated protein, demonstrating that PEF did not hydrolyze alkaline phosphatase. ALP activity was reduced by 44% after twenty 0.78 msec pulses of 22.3 kV/cm, with a concomitant increase in 90° light scattering, increased intrinsic fluorescence, and a blue shift of the intrinsic fluorescence emission spectra. The changes in fluorescence intensity and wavelength are consistent with PEF induced ALP aggregation that shields tryptophan residues from the polar solvent. The fluorescent intensity of the extrinsic CPA probe increased when seventy 0.78 msec pulses of 22.3 kV/cm were applied, but the fluorescence intensity was constant or decreased slightly for the extrinsic ANS probe. Given that CPA is more sensitive to aliphatic hydrophobicity, and

ANS to aromatic hydrophobicity, the extrinsic fluorescence probes indicate that PEF preferentially increases accessibility of aliphatic hydrophobic regions of alkaline phosphatase. A tentative model to explain the mechanism of PEF denaturation of protein consistent with the preliminary fluorescence data is presented and discussed.

INTRODUCTION

Pulses of high intensity electric fields (PEF) are a potentially important nonthermal pasteurization/stabilization food preservation technique that will replace or partially substitute thermal processing technologies (Castro et al., 1993). Inactivation of microorganisms in liquids is effectively performed by PEF. Bacteria, yeast, and mold cells in milk, yogurt, orange juice, and fluid eggs are reduced five log cycles by PEF (Castro et al., 1993). Alkaline phosphatase (ALP) is 65% inactivated when an electric field of 18.8 kV/cm is applied to ALP in a modified simulated milk ultrafiltrate and in milk (Castro et al., 1994).

The presence of alkaline phosphatase in milk was described 62 years ago. It is recognized that ALP in milk is sensitive to thermal denaturation. This finding was employed to distinguish raw milk, rich in ALP, from pasteurized milk, which contains very little enzymatic activity. The activity of ALP in pasteurized milk products is of public health significance, since the active presence of ALP indicates inadequate pasteurization or contamination of pasteurized milk with raw milk. In fresh raw milk, ALP is present in cream associated with the membranes of fat globules, and in skim milk in the form of lipoprotein particles (Morton, 1954).

Several research methods were employed in these investigations, including UV absorbance and turbidity, intrinsic tryptophan fluorescence, extrinsic fluorescence rising from two fluorescent probes, and polyacrylamide gel electrophoresis to investigate the mechanism of ALP denaturation caused by PEF.

Fluorescence spectroscopy is a sensitive and versatile optical technique for studying protein structure, dynamics, and interaction of proteins in solution (Eftink, 1991, 1994). Fluorescence of proteins can be either intrinsic or extrinsic. The intrinsic fluorescence of most proteins is due to the presence of the fluorescent aromatic amino acids tryptophan and tyrosine (Lackowicz, 1983; Bezkorovainy, 1986; Eftink, 1991, 1994). Extrinsic fluorescence arises from probes that undergo changes in one or more fluorescence properties as a result of noncovalent interaction with proteins (Stryer, 1965, 1968; Semisotnov et al., 1991). Two examples of molecules used as fluorescent probes are 1-anilinonaphthalene-8-sulfonate (ANS) (Akita and Nakai, 1990; Semisotnov et al., 1991) and cis-parinaric acid (CPA) (Kato and Nakai, 1980; Akita and Nakai, 1990; Duxbury et al., 1991). It is suggested that ANS and CPA distinguish between exposed aliphatic side chains and exposed intrinsic aromatic rings (Hayakawa and Nakai, 1985).

Molecular weights of proteins are determined by light scattering measurements. In addition to the molecular weight estimates, light scattering can also yield information on the shape and dimensions of protein molecules (Bezkorovainy, 1986; Wyatt, 1993). A simple indication of light scattering is the turbidity encountered in absorbance measurements (Demchenko, 1986).

In this study, direct evidence suggests that pulsed high intensity electric fields cause changes in the structure of the proteins, as revealed by changes in intrinsic and extrinsic fluorescence and turbidity. The differential behavior of CPA as compared to ANS suggests that the aliphatic hydrophobic groups are preferentially exposed on the surface of the PEF-denatured protein molecule, as compared to the aromatic hydrophobic groups. A simple model to account for these observations is that the PEF-induced surface change causes a compression of the protein, which leads to exposure of hydrophobic groups at the protein surface, which in turn facilitates protein-protein aggregation.

MATERIALS AND METHODS

PULSED ELECTRIC FIELDS

Alkaline phosphatase type XXT from bovine milk, 3.5 units/mg, was purchased from Sigma Chemical Co. (St. Louis, MO) and dissolved at 2 mg/mL in raw milk, 2% milk, nonfat milk, or modified simulated milk ultrafiltrate (MSMU) (Jenness and Koops, 1962).

The alkaline phosphatase dissolved in MSMUF was transferred to a sterile and disposable cuvette with a 0.1 cm interelectrode distance (BIO-RAD Lab., Hercules, CA). High intensity electric field pulses were applied with a GeneZapper (Kodak, New Haven, CT) capable of generating electric field intensities up to 2500 V when using a 7 μfarads capacitor. A 5 E2 resistor was placed in series with the cuvette to increase the electric field intensity. An exponential decay waveform pulse was produced by discharge of the capacitor. The electric field intensity declined over time as a function of the resistance in the circuit and the size of the capacitor. The exponential decay curve is described by:

$$E_{(t)} = E_0 e^{-t/\tau} \tag{1}$$

where $E_{(t)}$ is the electric field intensity (kVcm^{-1}) at any time t (s). E_0 is the initial field intensity, and τ is the resistance capacitance (RC) time constant (s). The time constant depends on the total resistance (R in Ω) and capacitance (C in F) of the system as follows:

$$\tau = RC \tag{2}$$

Therefore, τ describes the shape of the decay waveform, and is the time required for the electric field intensity to decline 37% of the initial value. Because the resistance of the solution affects the time constant, the composition of the solution and the geometry of the cuvette influence the shape of the discharge pulses. The resistance of the solution is inversely proportional to the ionic strength, temperature, and cross-sectional area of the solution-electrode interface. The resistance is directly proportional to the distance between the electrodes.

A maximum of 70 exponential decay pulses of 0.40 milliseconds for milk and 0.74 milliseconds for MSMUF were applied. Time between pulses was 15 s. The electric field intensity selected was 22.0 kVcm^{-1}. A parallel resistance of 100 ohms and a capacitance of 7 μfarads were selected for each PEF treatment.

ABSORBANCE AND TURBIDITY

One milliliter of untreated or PEF treated ALP was placed into a microcell with an inside width of 2.0 mm (Hellma Cells Inc., Jamaica, NY). The absorbance was measured in a diode array spectrophotometer (Hewlett Packard, 8425A, Waldbronn, FRG). The turbidity value was determined by plotting a graph of the log absorbance vs. log λ (Winder and Gent, 1971), and extrapolation of the straight line drawn at the 310–350 nm region, in which no absorption of chromophores occurs, into the wavelength region in which protein absorption occurs. After conversion of log A to A and subtraction from the observed absorption of the untreated and PEF treated ALP, the absorption spectra corrected for scattering was plotted (Donovan, 1969).

INTRINSIC FLUORESCENCE

Fluorescence spectroscopy is an established, widely used method for studying the structure, dynamics, and interactions of proteins in solution (Eftink, 1991). The usefulness of fluorescence spectroscopy is due to the rich variety of molecular details that it reveals about proteins, including solvent exposure of amino acid side chains, the existence of protein conformers, motion of the fluorophore or motion of molecules and protein side chains in the neighborhood of the fluorophore. The information obtained, together with the sensitivity of the measurements, the existence of intrinsic fluorescent group, and the ability to introduce specific extrinsic fluorescent groups, makes fluorescence spectroscopy a method of particular importance in studies of protein structural changes.

Fluorescence emission spectra of alkaline phosphatase were obtained by varying wavelength emission, while recording the intensity of excitation at a wavelength of 295 nm in an J4-8961 Aminco-Bowman spectrophotofluorometer (Silver Spring, MA). One millimeter and 2 mm slits were selected for the excitation exit slit and emission entrance slit, respectively. A rotatory slit of 2 mm was selected. The scale of 100% and the sensitivity of 0 were selected in the photomultiplier microphotometer.

EXTRINSIC FLUORESCENCE

Protein hydrophobicity was determined using CPA and ANS as fluorescence probes (Sklar et al., 1977). *cis*-Parinaric acid (Molecular Probes, Eugene, OR) was dissolved in absolute ethanol to obtain 3.6×10^{-3} M solution. The ethanolic solution of CPA contained butylated hydroxytoluene (BHT) as an antioxidant. Ten microliters of CPA (Kato and Nakai, 1980) solution were added to 2 mL of 2 mg/mL of ALP solution in MSMUF, pH 8.0. The CPA protein conjugates were excited at 325 nm and the relative fluorescence intensity was assayed at 420 nm in a J4-8961 Aminco-Bowman spectrophotofluorometer (Silver Spring, MA). One milliliter of 250 mM ANS in 0.1 M phosphate buffer, pH 6.8, was added to 1 mL of 2 mg/mL of ALP solution in modified MSMUF. The excitation wavelength was 350 nm.

ELECTROPHORESIS

Polyacrylamide gel electrophoresis (PAGE) under non-denaturing conditions was performed with the BioRad Proteum III System (BioRad Laboratories, Richmond, CA), using 12.0% separating acrylamide gels. The buffer was 62.5 mM Tris-HCL, pH 6.8, 10% glycerol, 2% SDS, and 5% *P*-mercaptoethanol. Alkaline phosphatase was located on the gels by means of 5-bromo-4-chloro-3-indolyl phosphate as the calorimetric substrate for ALP and detection of the orthophosphate with nitroblue tetrazolium (phosphate conjugated substrate, BioRad Laboratories, Richmond, CA).

MSMUF

MSMUF is a formulated system of milk constituents used in nonthermal processing research to study the properties of milk proteins and potential changes during nonthermal processing. Simulated milk ultrafiltrate (SN4UF) is a salt solution with a selected composition encountered in milk ultrafiltrates (Jenness and Koops, 1962). Lactose (5%) was added to depress the electrical conductivity of the simulated milk ultrafiltrate (MSMUF) to 500–600 μmhos/cm. MSMUF was adjusted to the optimum pH of 8.0 for ALP activity. MSMUF was passed through filters of 0.22 μm pore width (Falcon, Becton Dickinson Labware, Lincoln Park, NJ) to remove insoluble and denatured materials.

RESULTS AND DISCUSSION

POLYACRYLAMIDE GEL ELECTROPHORESIS (PAGE)

Previously, extensive treatment of milk with PEF led to 65% inactivation of ALP activity (Castro et al., 1994). Although a field induced conformational change in protein of some type seems most likely to account for the inactivation, for the sake of completeness, a test for field-induced chemical changes was conducted.

SDS-PAGE of ALP, treated with 50 pulses at 20.8 kV/cm, with pulse length of 0.70 msec of high intensity electric fields, and untreated ALP exhibited essentially identical protein bands. If the 50% loss of enzyme activity was due to peptide cleavage, for example, then the protein band from PEF treated ALP would be less intense than the protein band from untreated ALP. The enzyme activity dependent stain used in this experiment was reported to stain SDS-denatured ALP (Bingham and Malin, 1992). Evidently, dilution of SDS during the staining procedure allows the ALP to refold to a native active conformation. Since the protein band from PEF treated ALP exhibited an intensity similar to the protein band of the untreated control, the treated and untreated ALP exhibit similar renaturation following SDS treatment. The lack of protein hydrolysis argues against chemical loss of ALP activity, and suggests a PEF induced conformational change reversed by the action of SDS.

UV ABSORBANCE

Chromophores, tryptophan and tyrosin, buried in the interior of ALP, become exposed to solvent when ALP is treated with PEF. The change in ultraviolet absorption of ALP is a measure of denaturation (Eftink, 1991) of ALP by PEF. When the number of pulses increased, the absorbance of the ALP treated with 0.78 pulses of 22.3 kV/cm of high intensity electric fields increased (Figure 6.1), meaning that more chromophores were exposed to solvent each time the number of pulses was increased.

The absorbance of ALP treated with 22.3 kV/cm of PEF increases as the number of pulses increases. The maximum absorbance of untreated ALP was 0.167 at a wavelength of 278 nm. After 10, 20, 30, 50, and 70 pulses of 22.3 kV/cm of high intensity electric field absorbance increased from 0.167 to 0.168, 0.209, 0.260, 0.410, and 0.640, respectively. Untreated ALP exhibited a maximum absorbance at 278 nm. A blue shift from 278 to 272 run of the maximum absorbance was observed after PEF treatment of 70 pulses of 22.3 kV/cm. Ten pulses of 22.3 kV/cm did not change the maximum absorbance of ALP. Alkaline phosphatase treated with 20 and 30 pulses of 22.3 kV/cm exhibited maximum

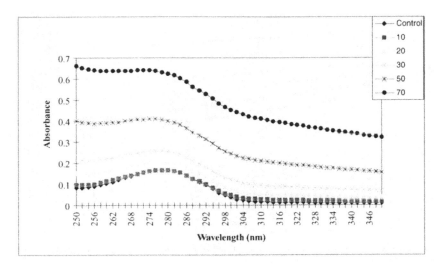

Figure 6.1 Absorption spectra of alkaline phosphatase treated with 0.78 msec pulses of 22.3 kV/cm of high intensity electric fields.

absorbance at 276 nm, and ALP treated with 50 pulses of 22.3 kV/cm exhibited maximum absorbance at 274 nm.

The measurable protein absorption spectra often have a component not associated with specific absorption bands, but determined by light scattering. Turbidity distorts spectra and makes obtaining the quantitative information difficult. The ALP turbidity produced by PEF was observed visually, and an objective measure estimated by light absorption in the region where specific absorption is absent, usually 330 nm for a protein solution (Figure 6.2). The

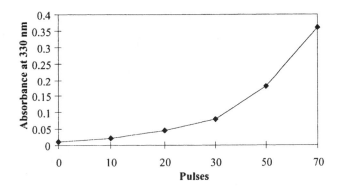

Figure 6.2 Turbidity of alkaline phosphatase treated with 22.3 kV/cm pulses of high intensity electric fields estimated by UV absorption at 330 nm.

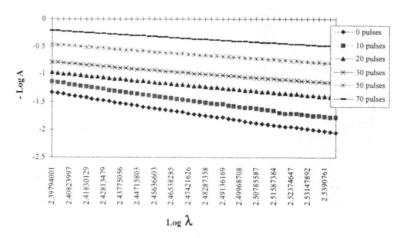

Figure 6.3 Double logarithmic plot of absorption spectra between 310 and 350 nm for ALP treated with 22.3 kV/cm pulses of high intensity electric fields.

main contribution to the turbidity of PEF treated ALP in solutions belongs to aggregates of ALP. Turbidity obtained by UV absorption was related to the visual observation of the increment of ALP turbidity after PEF treatment. The extrapolation of a straight line drawn at 310 to 350 nm, the wavelength region in which protein absorption occurred, is presented in Figure 6.3. The spectrum corrected for scattering is plotted in Figure 6.4. The increase of UV absorption at 278 nm by increasing the number of 0.78 msec pulses of PEF is presented in Figure 6.5.

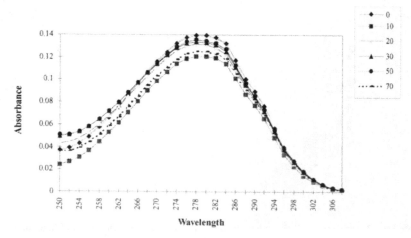

Figure 6.4 Absorption spectra of ALP treated with 22.3 kV/cm pulses of high intensity electric fields. The absorption spectra were corrected for turbidity.

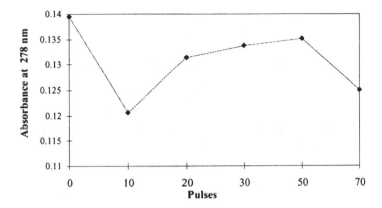

Figure 6.5 UV absorbance at 278 nm of ALP treated with 22.3 kV/cm pulses of high intensity electric fields as a function of number of pulses.

The results reported were obtained by using a GeneZapper as a source of electric pulse that produces a maximum electric field strength of 20.0–23.0 kV/cm and a pulse duration of 700–800 μsec.

Experiments were also carried out using the pulsed electric fields equipment developed at Washington State University (WSU). The duration of the electric field pulse when using the WSU equipment was 15–17 μsec, and the electrical field intensity was between 20 and 35 kV/cm. Fifty pulses of electric fields using WSU equipment did not inactivate ALP or cause turbidity of the ALP solution. Pulse duration is an important factor in the ALP inactivation effect of PEF. The loss of ALP activity correlated with protein aggregation.

UV absorption spectra of proteins provide information not only on electronic structure of chromophore groups, but also on the interaction of chromophores with the solvent environment (Demchenko, 1986). When tryptophan, tyrosine, and phenylalanine are transferred to a solvent environment with lower refractive index than the refractive index of the nonpolar interior region, a blue shift is observed in the UV spectrum (Bigelow and Gerschwind, 1960, cited in Demchemko, 1986). The blue shift of the UV absorption spectrum of ALP treated with 22.3 kV/cm pulses of high intensity electric fields (Figure 6.3) permits a simple explanation for ALP denaturation: the blue shift is associated with transition of chromophores into the MSMUF with a smaller value of the refractive index than the refractive index of the nonpolar interior region. Blue shift of the UV absorption spectrum also is associated with an increase of the dipole-dipole interactions (Wetlaufer, 1962). The degree of the UV absorption spectrum change depends on the distribution of the solvent molecules around proteins (Demchenko, 1986).

Tyrosine (Tyr) and tryptophan (Trp) contain groups of atoms, such as the OH group, capable of participating in hydrogen bonding. In the formation of hydrogen bonds, Tyr and Trp may act as proton donors or proton acceptors. A blue shift of the protein absorption spectrum is induced when the OH group from Tyr or Trp is a proton acceptor of the hydrogen bond (Kamlet et al., 1981). A blue shift is also produced with proton dissociation from the α-carboxylic group (Donovan, 1969). The change in polarization occurs parallel to denaturation of the chromophore environment in going from hydrophobic regions to hydrophilic aqueous environment, and is a primary factor in perturbation of the chromophore group spectra of the denatured protein. Protonation of carboxylic groups is of secondary importance (Donovan, 1969).

INTRINSIC FLUORESCENCE

Twenty pulses of 22.3 kV/cm with a pulse length of 0.78 msec of PEF produced the maximum increase of intrinsic fluorescence from 32.2 to 34.4 nm relative intensity, and a blue shift from 345 to 343 nm. Fifty pulses of 22.3 kV/cm produced a decrease in the intrinsic fluorescence from 32.2 to 27.7 nm relative intensity, and a blue shift from 345 to 341 nm (Figure 6.6). The relative intensity light scattering of ALP treated with PEF increased with increasing number of pulses. Relative intensity at 345 nm increased from 4 to 13, 18, 19, 24, and 49 when 10, 20, 30, 50, and 70 pulses of 22.3 kV/cm with pulse length of 0.78 msec were applied, respectively (Figure 6.7). The increase in relative intensity light scattering of bovine milk ALP occurred parallel to a reduction of enzymatic activity of 44% (Castro et al., 1994).

The increase in fluorescent relative intensity with increasing number of pulses is attributable to Trp shifting into a more nonpolar environment (Eftink, 1991). The shift from 345 to 343 nm may be due to a protein conformational change

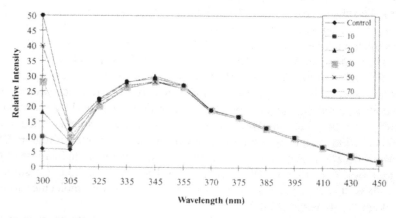

Figure 6.6 Fluorescence emission spectra of alkaline phosphatase treated with 22.3 kV/cm pulses of high intensity electric fields.

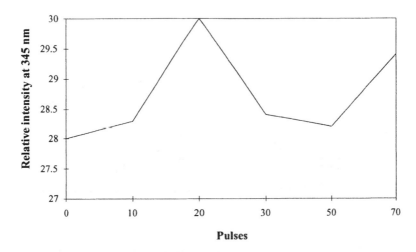

Figure 6.7 Fluorescence emission at 345 nm of ALP treated with 22.3 kV/cm pulses of high intensity electric fields as a function of number of pulses.

or to aggregation. Light scattering also increased with the increase in number of pulses. Fluorescent maximum (λ_{max}) of Trp residues in native proteins commonly ranges from about 325 to 350 nm. There is a relationship between the λ_{max} and the exposure of Trp in protein to an aqueous solvent (Burstein et al., 1973; Eftink and Ghiron, 1976). Surface amino acid residues exposed to water will fluoresce red (350 nm) and amino acid residues buried in a relative hydrophobic interior region of a protein will fluoresce blue (325 nm). Alternatively, a. Trp may fluoresce blue when sandwiched in a rigid portion of a protein, even if the environment is not particularly hydrophobic (Eftink and Ghiron, 1976).

Tryptophan fluorescent maximum emission of ALP treated with 22.3 kV/cm occurred at a shorter wavelength (blue shift) relative to the emission of Trp from untreated ALP. The blue shift is a result of the shielding of the Trp residues from a polar environment. After PEF treatment ALP aggregates were observed in the cuvette as dilute turbidity and confirmed by a small increase in UV light absorbance at 330 mn (Figure 6.2).

EXTRINSIC FLUORESCENCE

Fluorescent relative intensity of the extrinsic CPA probe bound to untreated ALP increased from 5.15 to 12.3 when CPA was bound to ALP treated with 20 pulses of 22.3 kV/cm. The fluorescent relative intensity of untreated ALP increased from 0.62 to 5.12 when bound to CPA (Figure 6.8). The fluorescent relative intensity of untreated ALP bound to CPA increased from 5.1 to 5.2, 6.8, 8.7, 9.8, and 12.3, respectively, when 70, 50, 10, 30, and 20 pulses

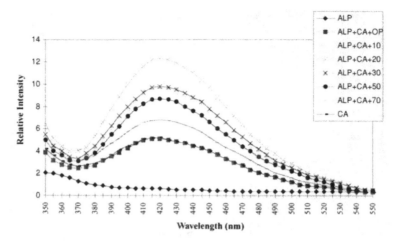

Figure 6.8 Fluorescence emission spectra of alkaline phosphatase treated with 0.78 msec pulses of 22.3 kV/cm of high intensity electric fields bound to *cis*-parinaric acid.

of 22.3 kV/cm were applied (Figure 6.9). A blue shift of the spectra of untreated ALP bound to CPA from 420 to 415 nm was observed after ALP was treated with 10 pulses of 22.3 kV/cm of high intensity electric fields and bound to CPA.

The native ALP structure is stabilized by a variety of interactions, including hydrogen bonds, hydrophobic interactions, and electrostatic interactions (Dill, 1990). Disruptions of side chain interactions by pulses of 22.3 kV/cm of high intensity electric fields altered the tertiary structure and denatured ALP. Concomitantly, many hydrophobic groups originally in the interior of the molecule

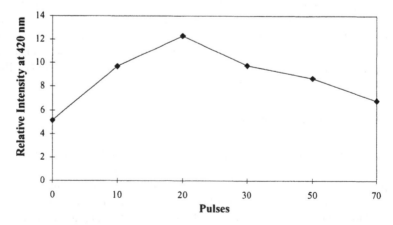

Figure 6.9 Fluorescence emission at 345 nm of ALP treated with 0.78 msec pulses of 22.3 kV/cm of high intensity electric fields and bound to *cis*-parinaric acid as a function of number of pulses.

are exposed to more polar solvent, MSMUF. The exposed hydrophobic groups interact in the hydrophilic environment, forming aggregates.

Hydrophobic fluorescent probes are used to investigate exposed hydrophobic sites in proteins (Stryer, 1965). Noncovalent binding of a fluorescent probe to a protein leads to changes in the fluorescence of the probe. Fluorescent relative intensity of ALP treated with pulses at 22.3 kV/cm of high intensity electric fields increased and shifted to a shorter wavelength (Figure 6.8). CPA and ANS probes were used in the present study to investigate exposed hydrophobic groups after PEF treatment of ALP. The affinity of hydrophobic probes to proteins increases when the rigidity of the ALP tertiary structure is disrupted, and the secondary structure is retained (Semisotnov et al., 1991). The ANS hydrophobic probe binds the hydrophobic region of proteins. The hydrophobic region must be sterically accessible and sufficiently extended to bind the probe (Stryer, 1965). With the formation of the hydrophobic aggregates, the hydrophobic region is not accessible to ANS, explaining why the λ_{max} did not change when ANS was added to ALP treated with PEF. If ALP were totally unfolded by pulsed electric fields, no binding of ANS to treated ALP would occur (Semisotnov et al., 1991).

A molten globule is an intermediate protein structure between native and denatured protein forms, that does not change in compactness (Semisotnov et al., 1991). Molten globules are characterized by a native-like secondary structure with a compact shape, but are distinguished from native proteins by a nonrigid side-chain arrangement and nonfixed tertiary structures (Ohgushi and Wada, 1983). The affinity of molten globules to the ANS probe is due to the absence of a rigid hydrophobic cluster (Semisotnov et al., 1991). Fluorescent relative intensity of alkaline phosphatase treated with 22.3 kV/cm pulses of high intensity electric fields bound to ANS did not change or increase. Therefore, ALP treated with PEF did not form a molten globule.

Hayakawa and Nakai (1985) classified the hydrophobicity of proteins as aliphatic, due to exposure of aliphatic amino acid residues to an aqueous environment, or aromatic, due to exposure of aromatic amino acid residues to an aqueous environment. Hayakawa and Nakai (1985) reported that CPA is composed of one aliphatic hydrocarbon chain and is useful for determining aliphatic hydrophobicity. ANS is composed of one aromatic ring and is more useful for determining aromatic hydrophobicity. Variation in the size, shape, or composition of the hydrophobic sites of select proteins during denaturation will affect protein interaction with probes, resulting in differences in probe fluorescence (Edelman and McClure, 1968). Treatment of proteins with pulses of high intensity electric fields rearranges the protein exposing aliphatic hydrophobic groups to the surface. The CPA intrinsic probe assay demonstrates that PEF increases the accessibility of aliphatic hydrophobic regions of the ALP. Altering of ALP conformation is accompanied by loss of biological activity (Castro et al., 1994).

An increase in the fluorescent relative intensity of alkaline phosphatase treated with 22.3 kV/cm pulses of PEF and bound to CPA (Figure 6.8) suggests that more aliphatic amino acids were exposed to an aqueous environment

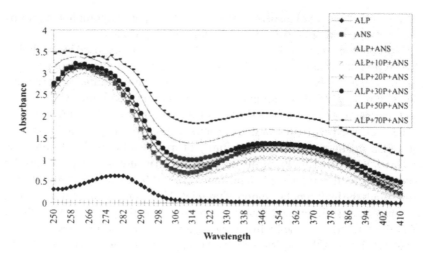

Figure 6.10 Absorption spectra of alkaline phosphatase treated with 0.78 msec pulses of 22.3 kV/cm of high intensity electric fields bound to ANS.

after PEF treatment, while exposure to an aqueous environment of aromatic amino acids changed little. The fluorescent relative intensity of alkaline phosphatase treated with 22.3 kV/cm pulses of high intensity electric fields bound to ANS remains constant or decreases slightly.

No change of ALP treated with 22.3 kV/cm of PEF bound to ANS was noticed at λ_{max} of 520 nm. Stryer (1968) emphasized that the fluorescence of protein chromphores is more sensitive to environment than is protein absorptivity. Demchenko (1986) reported that protein conformational changes, rather than shielding of chromophore groups, are responsible for changes in the protein absorption spectra. These two reports may explain why no change in the fluorescence spectrum was observed after ALP was treated, and why (Figure 6.10) the absorption spectrum of ALP treated with PEF and bound to ANS presented a blue shift and an increase in absorbance.

The absorption spectra of ALP treated with 22.3 kV/cm pulses of high intensity electric fields bound to CPA is shown in Figure 6.11. The spectrum of ALP treated with 20 pulses is closest to the spectrum of CPA, indicating that almost all of the CPA was bound to ALP treated with 20 pulses. The aliphatic hydrophobic residues of ALP treated with 20 pulses were exposed to a greater extent than the aliphatic hydrophobic residues of untreated ALP or ALP treated with 10, 30, 50, and 70 pulses of 22.3 kV/cm of PEF.

The absorbance of untreated ALP and ALP treated with 22.3 kV/cm of high intensity electric fields increased upon binding of CPA (Figure 6.11). The absorption spectra of ALP treated with PEF and bound to CPA exhibited three peaks at 278, 306, and 322 nm. The maximum absorbance of treated ALP shifted

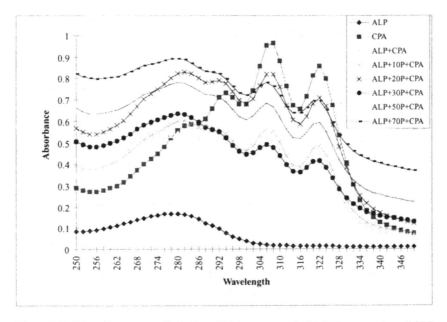

Figure 6.11 Absorption spectra of alkaline phosphatase treated with 0.78 msec pulses of 22.3 kV/cm of high intensity electric fields bound to *cis*-parinaric acid.

from 278 to 282 nm after binding to CPA. After 10 and 20 pulses at 22.3 kV/cm the maximum absorbance of treated ALP was similar to untreated ALP bound to CPA. After treatment of ALP with 30, 50, and 70 pulses at 22.3 kV/cm, the maximum absorbance shifted from 278 to 280 nm. ALP treated with 20 and 70 pulses at 22.3 kV/cm and bound to CPA exhibited maximum absorbance at 282 and 280 nm, respectively (Table 6.1). The maximum absorbance at 306 and 322 nm was observed (Table 6.1) when ALP was treated with 20 pulses at

TABLE 6.1. Absorption Spectrum Peaks ALP, Treated with 0.78 msec Pulses of 22.3 kV/cm of High Intensity Electric Fields, Bound to CPA.

	Peak 1		Peak 2		Peak 3	
	λ	Absorbance	λ	Absorbance	λ	Absorbance
ALP	278	0.167				
CPA			308	0.964	322	0.857
ALP + 0 P + CPA	282	0.548	306	0.527	322	0.456
ALP + 10 P + CPA	282	0.602	306	0.566	322	0.486
ALP + 20 P + CPA	282	0.83	306	0.820	322	0.708
ALP + 30 P + CPA	280	0.634	306	0.492	322	0.415
ALP + 50 P + CPA	280	0.779	306	0.682	322	0.590
ALP + 70 P + CPA	280	0.893	306	0.781	320	0.690

TABLE 6.2. Absorption Spectrum Peaks of ALP, Treated with 0.78 msec Pulses of 22.3 kV/cm of High Intensity Electric Fields, Bound to ANS.

	Peak 1		Peak 2	
	λ	Absorbance	λ	Absorbance
ALP	278	0.63		
ANS	262	3.14	350	1.37
ALP + 0 P + ANS	268	2.83	350	0.81
ALP + 10 P + ANS	264	2.99	350	1.07
ALP + 20 P + ANS	262	3.14	348	1.26
ALP + 30 P + ANS	260	3.21	348	1.39
ALP + 50 P + ANS	260	3.37	348	1.72
ALP + 70 P + ANS	256	3.52	346	2.09

22.3 kV/cm and reacted with CPA. ALP treated with 20 pulses at 22.3 kV/cm bound the greater amount of CPA, meaning 20 pulses at 22.3 kV/cm of PEF expose more aliphatic amino acids to the CPA than the other treatments.

UV maximum absorbance of treated ALP at 278 nm and ALP bound to ANS at 350 nm was shifted to a shorter wavelength (Table 6.2). Upon binding the ANS extrinsic probe, the maximum absorbance of untreated alkaline phosphatase increased from 0.63 to 2.83 and the wavelength shifted from 278 to 268 nm. Absorbance of alkaline phosphatase treated with 22.3 kV/cm bound to ANS increased from 1.07 to 2.09, and the wavelength shifted from 350 to 346 nm. Seventy pulses at 22.3 kV/cm of high intensity electric fields resulted in greater absorbance and a shift to a shorter wavelength of treated ALP bound to ANS (Table 6.2).

THEORY AND MODEL

ALP molecules treated with high intensity electric fields of 22.3 kV/cm with a pulse length of 0.78 msec tend to associate, forming aggregates. The aggregate formation was confirmed by visual observation, light scattering, and turbidity as a measure of UV absorbance at a wavelength (λ) of 330 mn. Pulses on the order of μseconds did not cause aggregation of ALP. No turbidity was observed after the PEF treatment using the equipment from the Washington State University electric processing technique because the pulse length is on the order of μseconds. The model proposed to explain the effects of high intensity pulsed electric fields on proteins is illustrated in Figure 6.12. The native molecule of ALP (A) is polarized by PEF (B). After applying more pulses, aggregates of ALP are formed (C and D). Protein aggregates are formed by the electrostatic attraction created by electrical charges of the dipole on the same protein (Schelly and Astumian, 1984).

Aliphatic hydrophobic interaction plays a role in the formation of the aggregates. PEF caused conformational changes in ALP. The aliphatic amino

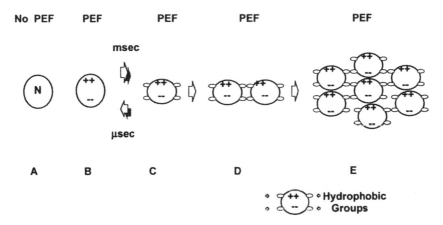

Figure 6.12 Proposed model to explain the effects of PEF on proteins.

acids are buried inside the native ALP, making an important contribution to the maintenance of the native globular structure of the protein (Mozhaev and Martinek, 1984). Some aliphatic amino acids of ALP are transferred from the hydrophobic interior onto the hydrophylic surface of the ALP molecule when treated with PEF. The released aliphatic amino acids are characterized when the relative intensity of the fluorescent spectra of PEF treated ALP bound to *cis*-parinaric acid (CPA) increased (Hayakawa and Nakai, 1985). No changes were observed in the fluorescent spectra when an aromatic hydrophobic probe, ANS, was used. Tryptophan and tyrosine, the aromatic amino acids responsible for the fluorescence properties of proteins, are not present in the region of intramolecular contact, or in the formation of associates (Semisotnov et al., 1991), so no conformational changes on ALP treated with PEF influence fluorescent spectra.

A peptide unit exhibits electric dipole of ~3.46 Deby (D) (Hol et al., 1978). In α-helical proteins peptide dipoles are aligned in the same direction, i.e., the positive pole of the dipoles points to the N-terminus of the peptide, and the negative pole of the dipole points to the C-terminus of the peptide (Hol et al., 1978) (Figure 6.13). Thus, an α-helix composed of n amino acids is a dipole of $3.46n$ D. β-Sheets, on the other hand, exhibit little net dipole moment (Hol et al., 1981). Pulses of high intensity electric fields will change the conformation of the protein producing a larger dipole (Schelly and Astumian, 1984). Dipoles aggregate because of PEF, and during aggregation the strength of the dipole decreased because of partial neutralization at the poles.

Three important effects of electric fields on molecules involve: (1) movement of free electrons, ions, and other charged particles; (2) polarization, e.g., the displacement of bound charges, electrons in atoms, atoms in molecules; (3) orientation of the molecular dipole moment (Presman, 1970; De Maeyer, 1969); and (4) a change in the dielectric constant of molecules.

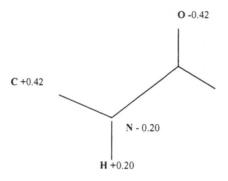

O -0.42

C +0.42

N - 0.20

H +0.20

Figure 6.13 Geometry and dipole moment of the peptide unit. The dipole moment has a value of 3.45 D (Hol et al., 1978).

Molecular orientation originated by the high intensity electric fields increases the dielectric constant (D) of the peptide unit for nonpolar interactions. The dielectric constant and dipole moments of α-helix tend to increase simultaneously. According to Coulomb's law:

$$F = \frac{kq_1q_2}{Dr^2}$$

where F is the force between two electrical charges in a dipole, q_1 and q_2, separated by the distance r; D is the dielectric constant of the medium between charges; and k is a proportionality constant. Consequently, at large D values, the forces between electric charges in a dipole will be much weaker than at small D values.

Edelman and McClure (1968) reported a close parallelism between wavelength of maximal emission values and the reduction of dielectric constant of the solvent such as water. Figures 6.6 and 6.7 present λ_{max} values when ALP was treated with 0.78 msec pulses of 22.3 kV/cm of high intensity electric fields. The maximum values were obtained when twenty 0.78 msec pulses of 22.3 kV/cm were applied to alkaline phosphatase in MSMUF. The λ_{max} values of ALP treated with 20 pulses of 22.3 V/cm correspond to the maximum λ_{max} values of D for ALP in MSMUF. Using Coulomb's law, the force (F) values of the ALP dipole will be the minimum producing denaturation of ALP by disruption of the dipole-dipole interaction forces.

Native ALP exhibits a helical conformation in which the polar residues align in one direction, positive charges at N-terminus of the helix and negative charges at the C-terminus parallel to the helical axis (Hol et al., 1981). PEF treatment increased the dielectric constant of ALP, resulting in a polypeptide chain with more internal freedom of motion. Dipole-dipole interaction plays an important, if not predominant, role in stabilizing protein α-helices.

The results of ALP exposure to the electric field, and changes in molecular structure can be summarized as follows: (a) polarization of the ALP; (b) polar structures tend to attract each other by electrostatic forces; (c) aliphatic hydrophobic amino acids are exposed to the aqueous surface of the protein; and (d) aliphatic hydrophobic amino acids interact and form aggregates producing an instability of the electrostatic forces that maintain the native structure, causing a reduction of the ALP activity and denaturation of the ALP protein (Castro et al., 1994).

CONCLUSIONS

Polyacrylamide gel electrophoresis of ALP treated with fifty 0.70 msec pulses of 20.8 kV/cm provided evidence that PEF did not cause proteolysis; and that the blue shift of fluorescence emission spectra, and the increase in fluorescent relative intensity observed, reflected subtle alterations in the ALP structure. Therefore, ALP was weakly denatured by PEF.

The decreae in fluorescent relative intensity of ALP treated with PEF showed PEF induced aggregation, and shielded tryptophan residues from MSMUF.

Aliphatic hydrophobic amino acids were exposed in greater proportion to the surface of the ALP treated with PEF. These amino acids bound *cis*-parinaric acid and increased the fluorescent relative intensity of the ALP treated with PEF and bound to CPA.

ALP treated with PEF denatured and formed aggregates that were noticed by visual observation, light scattering, and measured by UV absorption at 330 nm.

NOMENCLATURE

ALP   = alkaline phosphatase
ANS   = 1-anilinonaphthalene-8-sulfonate
CPA   = *cis*-parinaric acido
MSMUF = modified simulated milk ultrafiltrate
PAGE   = polyacrylamide gel electrophoresis
PEF   = pulses of high strength electric fields
E_0 = the initial field strength
$E_{(t)}$ = the electric field strength (V/cm) at any time t (s);
         V: voltage; t: time (s)
τ = the resistance capacitance (RC) time constant (s)
R = resistance
C = capacitance
F = farads
Ω = ohms

REFERENCES

Akita, E. M. and Nakai, S. 1990. Lipophilization of β-lactoglobulin: Effect on hydrophocity, conformation and surface properties. *J. Food Sci.* 55:711–717.

Bezkorovainy, A. 1986. *Basic Protein Chemistry.* Ch. C. Thomas Publisher, Springfield, IL. Chapter 11.

Bingham, E. W. and Malin, E. L. 1992. Alkaline phosphatase in the lactating bovine mammary gland and the milk fat globule membrane release by phosphatidylinositol-specific phospholipase C. *Comp. Biochem. Physiol.* Vol. 102B:213–218.

Burstein, E. A., Vedenkina, N. S., and Ivkova, M. N. 1973. Fluorescence and the location of tryptophan residues in protein molecules. *Photochem. Photobiol.* 18:263–279.

Castro, A. J. 1994. Pulsed electric field modification of activity and denaturation of alkaline phosphatase. Ph.D. thesis, Washington State University, Pullman, WA.

Castro, A. J., Barbosa-Cánovas, G. V., and Swanson, B. G. 1993. Microbial inactivation of foods by pulsed electric fields. *J. Food Proc. Pres.* 17:47–73.

De Maeyer, L. C. M. 1969. Electric field methods. In: *Methods Enzymology* Kustin, K., Ed. Academic Press, NY, Chapter 14, pp. 80–118.

Demchenko, A. P. 1986. Ultraviolet Spectroscopy of Proteins. Springer-Verlag, Berlin, Germany, pp. 241–261.

Dill, K. A. 1990. Dominant forces in protein folding. *Biochem.* 29:7133–7155.

Donovan, J. W. 1969. Ultraviolet absorption. In: *Physical Principal and Techniques of Protein Chemistry.* Academic Press, London, pp. 101–170.

Duxbury, C. L., Legge, R. L., Paliyath, G., and Thompson, I. E. 1991. Lipid breakdown in smooth microsomal membranes from bean cotyledons alters membrane proteins and induces proteolysis. *J. Exp. Bot.* 42:103–112.

Edelman, G. M. and McClure, W. O. 1968. Fluorescent probes and conformation of proteins. *Accounts Chem. Res.* 1:65–70.

Eftink, M. R. 1991. Fluorescence techniques for studying protein structure. *Methods Biochem. Anal.* 35:127–205.

Eftink, M. R. 1994. The use of fluorescence methods to monitor unfolding transitions in proteins. *Biophys. J.* 66:482–501.

Eftink, M. R. and Ghiron, C. A. 1976. Exposure of tryptophanyl residues in proteins. Quantitative determination by fluorescence quenching studies. *Biochem.* 15:672–680.

Hayakawa, S. and Nakai, S. 1985. Relationships of hydrophobicity and net charge to the solubility of milk and soy proteins. *J. Food Sci.* 50:486–491.

Hol, W. G. J., van Duijen, P. T., and Berendsen, H. J. C. 1978. The α-helix dipole and the properties of proteins. *Nature.* 273:443–446.

Hol, W. G. J., Halie, L. M., and Sander, C. 1981. Dipoles of the X-helix and 0 sheet: Their role in protein folding. *Nature.* 294:532–536.

Jenness, R. and Koops, J. 1962. Preparation and properties of salt solution which simulates milk ultrafiltrate. *Neth. Milk Dairy J.* 16:153–164.

Kamlet, M. J., Dickson, C., and Taft, R. W. 1981. Linear salvation energy relationships. Solvent effects on some fluorescent probes. *Chem. Phys. Lett.* 77:69–72.

Kato, A. and Nakai, S. 1980. Hydrophobicity determined by a fluorescence probe method and its correlation with surface properties of proteins. *Biochim. Biophys. Acta.* 624:13–20.

Lakowicz, J. R. 1983. *Principles of Fluorescence Spectroscopy.* Plenum Press, NY, Chapter 11.

Morton, R. K. 1954. Some properties of alkaline phosphatase of cow's milk and calf intestinal mucosa. *Biochem. J.* 60:573–582.

Mozhaev, V. V. and Martinek, K. 1984. Structure-stability relationships in proteins: New approach to stabilizing enzymes. *Enz. Microb. Technol.* 6:50–59.

Ohgushi, M. and Wada, A. 1983. Molten-globule state: A compact form of globular proteins with mobile side-chains. *FEBS lett.* 164:21–24.

Presman, A. S. 1970. *Electromagnetic Fields and Life.* Plenum, New York, pp. 25–64.

Schelly, Z. A. and Astumian, R. D. 1984. A theory for the apparent "negative second Wien effect" observed in electric-field-jump studies of suspensions. *J. Phys. Chem.* 88:1152–1156.

Semisotnov, G. V., Radionova, N. A., Razgulyaev, O. I., and Uversky, V. N. 1991. Study of the molten globule intermediate state in protein folding by a hydrophobic fluorescent probe. *Biopolymers.* 31:119–128.

Sklar, L. A., Hudson, B. S., and Simoni, R. D. 1977. Conjugated polyene fatty acids as fluorescent probes: Binding to bovine serum albumin. *Biochemistry.* 16:5100–5108.

Stryer, L. 1965. The interaction of naphthalene dye with apomyoglobin and apohemoglobin. A fluorescent probe of non-polar binding sites. *J. Mol. Biol.* 13:482–495.

Stryer, L. 1968. Fluorescence spectroscopy of proteins. *Science.* 162:526–533.

Wetlaufer, D. B. 1962. Ultraviolet spectra of proteins and amino acids. *Adv. Protein Chem.* 17:303–390.

Winder, A. F. and Gent, L. G. 1971. Corrections of light scattering errors in spectrophotometric protein determination. *Biopolymers.* 10:1243–1251.

Wyatt, P. J. 1993. Light scattering and the absolute characterization of macromolecules. *Anal. Chim. Acta.* 272:1–40.

CHAPTER 7

Change in Susceptibility of Proteins to Proteolysis and the Inactivation of an Extracellular Protease from *Pseudomonas fluorescens* M3/6 When Exposed to Pulsed Electric Fields

H. VEGA-MERCADO
J. R. POWERS
O. MARTÍN-BELLOSO
L. LUEDECKE
G. V. BARBOSA-CÁNOVAS
B. G. SWANSON

ABSTRACT

A N extracellular protease from *Pseudomonas fluorescens* M3/6 was purified and concentrated using ultrafiltration. The protease had a molecular weight of 45–50 kDa and an optimum pH range of 6.0 to 8.0 at 37°C. Pulsed electric field (PEF) treatments on the protease were carried out in a tryptic soy broth enriched with a yeast extract (TSB/YE), skim milk, and casein-Tris buffer. Inactivation of the protease was a function of the intensity of the electric field, number of pulses, and medium containing the protease. An 80% inactivation was attained when a measured electric field of 18 kV/cm and 20 pulses at 0.25 Hz were used to treat the TSB/YE mixture. The inactivation changed significantly when the enzyme was treated in skim milk and exposed to measured electric fields of 14 and 15 kV/cm and pulsing rates of 1 and 2 Hz, when only 40% and 60% inactivation were achieved after 32 and 98 pulses, respectively. However, skim milk containing the protease and treated with a 25 kV/cm electric field at 0.6 Hz showed an increased proteolytic activity. Treating skim milk with PEF increased the susceptibility of the constituent proteins to the action of untreated protease. Experiments using casein-Tris buffer instead of skim milk demonstrated no significant inactivation of the protease, and no change in the susceptibility of casein to proteolysis.

105

INTRODUCTION

The storage of milk at 3°C to 6°C results in spoilage due to heat stable enzymes, protease and lipase, produced by psychrotrophic microorganisms that attack proteins and fat in the milk (Law, 1979; Cousin, 1982; Manji et al., 1986; Manji and Kakuda, 1988). Psychrotrophic bacteria are ubiquitous in nature and common contaminants of milk; the inactivation of psychrotrophic *Pseudomonas* is irreversible when milk is heated at 60°C for 30 min. Processing temperatures below 60°C allow *Pseudomonas* to recover and grow normally (Dabbah et al., 1969).

Microbial protease activity leads to bitter flavor development or coagulation of milk (Malik and Swanson, 1974; Law et al., 1977; Law, 1979; McKellar, 1981; Cousin, 1982; Manji et al., 1986; Manji and Kakuda, 1988). The proteases produced by *P. fluorescens* are thiol proteinases with an optimum pH of 7.0, if casein is used as substrate. In addition, *P. fluorescens* proteases are EDTA sensitive, contain Zn^{+2} (Barach et al., 1976b; Barach and Adams, 1977), and have molecular weights of 45,000 to 55,000 Da (Law, 1979).

The thermal resistance properties of proteases from *Pseudomonas* have been reported by Adams et al. (1975), Barach et al. (1976a, 1978), Barach and Adams (1977), Christen and Marshall (1984), Christen et al. (1986), and Carlez et al. (1993). Milk favors the activity of proteases from *Pseudomonas* that resist UHT treatment, because the pH is close to 7.0, which is optimum for proteolytic activity (Barach et al., 1978). The thermal stability of proteases from *Pseudomonas* is also related to structural flexibility and the interplay of divalent cations that allow enzyme renaturation after heating (Barach and Adams, 1977). Consequently, 56% to 60% inactivation of proteolytic enzymes from *Pseudomonas* via thermal methods requires up to 60 min at 60°C (Barach et al., 1976a), and is described in terms of the aggregation and autolysis of the protein (Barach et al., 1978).

The nonthermal inactivation of *Escherichia coli*, *Staphylococcus aureus*, *Allicrococcus lysodeikticus*, *Sarcina lutea*, *Bacillus subtilis*, *Bacillus megaterium*, *Clostridium welchii*, *Saccharomyces cerevisiae*, and *Candida utilis* using PEF was demonstrated by Sale and Hamilton (1967). In general, an increase in the electric field intensity and number of pulses leads to an increase in the inactivation of microorganisms (Jacob et al., 1981; Hülsheger et al., 1983; Sato et al., 1988; Mizuno and Hori, 1988; Pothakamury et al., 1996; and Zhang et al., 1994). Other factors that affect the microbial inactivation are treatment temperature, pH, ionic strength, and conductivity of the medium containing the microorganisms (Pothakamury et al., 1996).

Gilliland and Speck (1967) claimed the inactivation of trypsin, lactic dehydrogenase, and a protease from *B. subtilis* could be obtained with an electric field of 31.5 kV/cm and 20 pulses. Thus, they proposed that an inactivation mechanism for the enzyme is an oxidative reaction induced by the electric discharge as a function of treatment time. Castro (1994) arrived at the inactivation of alkaline phosphatase in SMUF by applying a PEF of 20 kV/cm, hence suggesting that conformational change on the tertiary structure of the protein in which hydrophobic groups within the alkaline phosphatase were realigned and caused aggregation and inactivation of the enzyme.

This study was designed to examine the inactivation of an extracellular protease from *P. fluorescens* M3/6 contained in tryptic soy broth, skim milk, and casein buffer by PEF, with the number of pulses, electric field strength, and total energy input after the PEF treatments as the factors under consideration.

MATERIALS AND METHODS

GROWTH OF *PSEUDOMONAS FLUORESCENS* M3/6, PROTEASE COLLECTION AND CHARACTERIZATION

The *Pseudomonas fluorescens* M3/6 strain was obtained from Dr. S. S. Nielsen, Food Science Department, Purdue University. Tryptic soy broth and agar (TSB, TSA) enriched with yeast extract (YE, 0.6% w/v) were used to culture the cells in a 28-day cycle to preserve the production of extracellular protease.

Sterile TSB/YE in volumes of 100 mL were inoculated with the *P. fluorescens* M3/6, incubated at 25°C for 24 hr, followed by an aging process at 4°C to 6°C for 18 days. Most of the *Pseudomonas* cells suspended in TSB/YE were removed by centrifugation at 4700 × g for 5 min followed by a sterile filtration step using a 0.45 μm filter. The filtered solution was then tested for proteolytic activity using the Pierce QuantiCleave Protease Assay Kit (Pierce, Rockford, IL).

The molecular weight of the chemical species contained in the sterile filtrate was narrowed by ultrafiltration using 50 kDa and 30 kDa membrane filters (MiCroSeP™, Filtron Technology Corp., Northborough, MA). The selection was based on a molecular weight of 45 kDa reported for the *P. fluorescens* M3/6 protease (Kohlman et al., 1991). The first step in the purification process was the removal of molecular species with a molecular weight greater than 50 kDa. Three mL of TSB/YE mixture were placed in the upper reservoir of the 50 kDa membrane filter, centrifuged at 4700 × g for 60 min at 0°C, then collected and transferred to the reservoir of a 30 kDa membrane filter for additional filtration and concentration by centrifugation at 4700 × g for 60 min at 0°C. The concentrated TSB/YE mixture containing the protease from the 30 kDa membrane filter was collected, filtered through a 0.45 μm sterile filter and stored at 4°C under sterile conditions. Two hundred mL of concentrated TSB/YE mixture with a protein content of 6.4 mg/mL were obtained from 600 mL of initial TSB/YE broth.

The optimum temperature for protease proteolytic activity in the TSB/YE mixture was determined with the Pierce QuantiCleave Protease Assay Kit using 0, 10, 23, 37, and 50°C as incubation temperatures for hydrolysis of the casein substrate with pH 6.2. Optimum pH was determined at the optimum temperature by adjusting the pH of the casein substrate solution (BIO-RAD, Richmond, CA) to 2.77, 4.33, 5.10, 6.20, 7.21, 8.50, 9.97, and 11.75 by titration with 1.0 M HCl or 1.0 M NaOH.

The molecular weight distribution in the TSB/YE mixture was determined using 4% to 20% Gradient Mini-Protean II Ready Gels (BIO-RAD, Richmond, CA), and electrophoresis was performed in a Mini-PROTEAN II cell (BIO-RAD, Richmond, CA) at a constant 200 V d.c. voltage for 30 min. The gel was stained with Coomassie Blue R-250 for 60 min and then destained with 10% acetic acid and 40% ethanol solution for 60 min or Copper staining as recommended by BIO-RAD (Richmond, CA). Molecular weight standards (BIO-RAD, Richmond, CA) containing myosin, β-galactosidase, phosphorylase b, bovine serum albumin, ovalbun-dn, carbonic anhydrase, soybean trypsin inhibitor, lysozyme, and aprotinin were used for the molecular weight determination.

ENZYMATIC AND PROTEIN ASSAYS

Bicinchoninic acid reagent (BCA*) was prepared by mixing 50 parts bicinchoninic acid and 1 part copper sulfate (Pierce, Rockford, IL); once mixed, 2 mL of the BCA* were dispensed into 10 mL glass vials and stored at room temperature for further use.

Proteolytic Assay

A 700 μL aliquot of the concentrated TSB/YE mixture containing the protease was combined with 2800 μL 1% casein solution provided by the QuantiCleave Protease Assay Kit (Pierce, Rockford, IL) and incubated at 37°C for 60 min. Two 500 μL samples were then removed at 0, 30, and 60 min for analysis. Sixty μL of 6.12 molal trichloroacetic acid (TCA) were added to each sample to precipitate the non-hydrolyzed casein. The solution was centrifuged at 10,000 × g and the supernatant filtered through a 0.45 μm membrane filter to remove the precipitated casein. A 200 μL aliquot from each TCA-supernatant sample was added to 2mL BCA* and incubated for 30 min at 37°C. Absorbance at 562 nm was measured using an HP 8452A diode array spectrophotometer (Hewlett Packard, Waldbronn, FRG). A blank of 1% casein and water solution was selected in each determination. Figure 7.1 summarizes the assay protocol.

A standard curve of bovine serum albumin (BSA) was selected as a reference protein. The proteolytic activity of the concentrated TSB/YE mixture containing

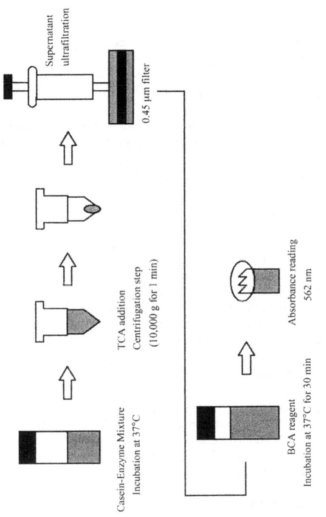

Figure 7.1 Proteolysis assay procedure.

Supernatant
ultrafiltration

0.45 μm filter

Casein-Enzyme Mixture
Incubation at 37°C

TCA addition
Centrifugaation step
(10,000 g for 1 min)

BCA reagent
Incubation at 37°C for 30 min

Absorbance reading
562 nm

the protease was expressed as specific activity (sp. ac.) and calculated as follows:

$$\text{sp. ac.} = \frac{\text{net change in solubilized protein } (\mu g)}{\mu g \text{ enzyme } - s}$$

ENZYMATIC INACTIVATION AND SUSCEPTIBILITY TO PROTEOLYSIS EXPERIMENTS FOR SKIM MILK AND CASEIN-TRIS BUFFER

Enzymatic Inactivation Experiments

The activity of the recovered protease from *P. fluorescens* M3/6 was determined before and after PEF treatments, while suspended in skim milk or a casein-Tris buffer. The effectiveness of the PEF treatments was expressed as the change in sp. ac. of the protease as a function of the PEF treatment conditions.

Susceptibility Experiments

Skim milk and the casein-Tris buffer were exposed to different PEF treatments before adding the protease from *P. fluorescens* M3/6. Once the substrates (skim milk and the casein buffer) were exposed to the PEF treatment, a predetermined amount of the protease was added into the solution and the sp. ac. of the protease from *P. fluorescens* M3/6 was determined. The susceptibility to proteolysis is expressed as the change in specific activity of the protease as a function of the PEF treatment conditions.

Total Protein Content

The BCA* reagent was selected to estimate the protein concentration in the TSB/YE mixture containing the protease from *P. fluorescens* M3/6. A 100 μL aliquot of the TSB/YE mixture was added to 2 mL of BCA* and incubated for 30 min at 37°C. The amount of protein in the TSB/YE mixture was determined from the BSA standard curve.

PEF EXPERIMENTS

The effect of PEF on the proteolytic activity of the protease from *P. fluorescens* M3/6 was studied considering three different media containing the protease: tryptic soy broth, skim milk, and a casein-Tris buffer solution. Meanwhile, the effect of PEF on the susceptibility to proteolysis of casein was studied using skim milk and a casein-Tris buffer solution. The description and details for each experiment follow.

TABLE 7.1. PEF Inactivation of Protease from *P. fluorescens* M3/6 in TSB/YE.

Measured electric field	11 kV/cm	18 kV/cm
Total energy input	6.1 J/pulse	20.3 J/pulse
Initial temperature	15°C	15°C
Final temperature	20°C	24°C
Number of pulses	0, 10, 20	0, 10, 20
Pulsing rate	0.25 Hz	0.25 Hz
Protein concentration*	200 μg/mL	200 μg/mL

*TSB/YE mixture containing the protease.

Test Solutions and PEF Treatment Conditions

Protease in Tryptic Soy Broth Enriched with 0.6% Yeast Extract (TSB/YE)

One L of TSB/YE with approximately 200 μg/mL of protein mixture containing the protease from *P. fluorescens* M3/6 was prepared for this experiment, with the percentage of inactivation evaluated as the reduction in sp. ac. of the protease suspended in the TSB/YE after the PEF treatments described in Table 7.1.

Sterile Skim Milk

Two 2 L batches of skim milk, containing 0.05% sodium azide to prevent the microbial spoilage of milk during the incubation period, were prepared as recommended by DIFCO (Detroit, MI) and sterilized at 121°C for 15 min. The percentage of inactivation was evaluated as the reduction in sp. ac. of the protease suspended in the skim milk after the PEF treatments (Table 7.2). The details for each experiment follow.

INACTIVATION OF THE PROTEASE INOCULATED INTO SKIM MILK

Ten mL (experiment 1) or 5 mL (experiment 2 and experiment 3) of TSB/YE mixture containing the protease from *P. fluorescens* M3/6 was added to 1 L of sterile skim milk five minutes before the PEF treatment, with the sp. ac. of the protease in skim milk assayed as a function of the selected number of pulses. Samples (12 mL) collected from the pulser were incubated at 37°C, and the proteolytic activity assayed using the BCA* method. The relative activity for the assayed samples was expressed as a percentage of specific activity remaining after the selected number of pulses. The activity of the protease on casein was determined by the BCA* assay method.

TABLE 7.2. **PEF Treatments of Skim Milk.**

Process Parameter	Experiment 1	Experiment 2	Experiment 3
Measured electric field	25 kV/cm	14 kV/cm	15 kV/cm
Total energy input	56 J/pulse	19 J/pulse	20 J/pulse
Initial temperature	17°C	14°C	20°C
Final temperature	30°C	30°C	50°C
Number of pulses	0, 8, 16	0, 16, 32	0, 35, 63, 98
Pulsing rate	0.6 Hz	1 Hz	2 Hz
Protein concentration*	64.1 µg/mL	32.1 µg/mL	32.1 µg/mL
Control samples (µg soluble protein/µg enzyme s)			
Inactivation tests	14.8 E-6	93.7 E-6	87.3 E-6
Susceptibility tests	15.2 E-6	84.1 E-6	

*TSB/YE mixture containing the protease.

SUSCEPTIBILITY OF CASEIN IN SKIM MILK TO PROTEOLYSIS DURING PEF TREATMENTS

Sterile skim milk was exposed to PEF treatments (Table 7.2, experiment 1 and 2) and 12 mL samples collected from the pulser. Immediately after sampling, 120 µL (experiment 1) or 60 µL (experiment 2) of the TSB/YE mixture containing the protease were added to the samples, and the proteolytic activity monitored as a function of time.

Sterile Casein-Tris Buffer

Two batches of casein-Tris buffer were prepared by mixing and boiling casein (10% and 2%, respectively) with 50 mM Tris buffer (pH 7.5), and sterilized at 121°C for 15 min. The first batch was used to conduct experiment 4, while the second batch, containing 0.05% w/v sodium azide to prevent microbial spoilage, was used in experiment 5.

INACTIVATION OF THE PROTEASE INOCULATED INTO CASEIN-TRIS BUFFER

Ten mL (experiment 4) or 5 mL (experiment 5) of TSB/YE mixture containing the protease from *P. fluorescens* M3/6 were added to 1 L of selected sterile casein-Tris buffer (10% or 2% w/v casein) five minutes before the PEF treatment. Samples (12 µL) collected from the pulser were incubated at 37°C and the proteolytic activity assayed using the BCA* method. The sp. ac. of the protease in the casein-Tris buffer was assayed as a function of the selected

TABLE 7.3. PEF Treatments of Casein-Tris Buffer.

Process Parameter	Experiment 4	Experiment 5
Measured electric field	32 kV/cm	23 kV/cm
Total energy input	90 J/pulse	49 J/pulse
Initial temperature	19°C	12°C
Final temperature	31.6°C	30.7°C
Number of pulses	16	32
Pulsing rate	0.6 Hz	1 Hz
Protein concentration*	64.1 µg/mL	32.1 µg/mL
Control samples (µg soluble protein/µg enzyme s)		
Inactivation tests	151.2 E-6	156.3 E-6
Susceptibility tests	124.7 E-6	103.4 E-6

*TSB/YE mixture containing the protease.

number of pulses, and expressed as the percentage ratio of the sp. ac. remaining after the selected number of pulses.

SUSCEPTIBILITY OF CASEIN IN A CASEIN-TRIS BUFFER TO PROTEOLYSIS DURING PEF TREATMENTS

Sterile casein-Tris buffer was exposed to PFF treatments (Table 7.3, experiment 4, and experiment 5) and 12 mL samples were collected from the pulser. Immediately after sampling, 120 µL (experiment 4) or 60 µL (experiment 5) of the TSB/YE mixture containing the protease was added to the samples and the proteolytic activity monitored as a function of time, while being incubated at 37°C.

High Voltage Repetitive Pulser

PEF treatments were carried out using a pilot plant size pulser (Figure 7.2), manufactured by Physics International (San Leandro, CA). A continuous treatment chamber, with a capacity of 28.5 mL and a 0.6 cm gap connected to a peristaltic pump to provide a constant flow rate of 0.5 L/min, was selected. The pulsing rate (duration of 2 µs with a 0.5 µF capacitor) was controlled by computer, and the voltage selected for the PEF treatments was set with a d.c. high voltage power supply.

STATISTICAL ANALYSIS

A completely random design was selected for the collection of experimental data, involving the mean of two replications with duplicate analyses for each.

114

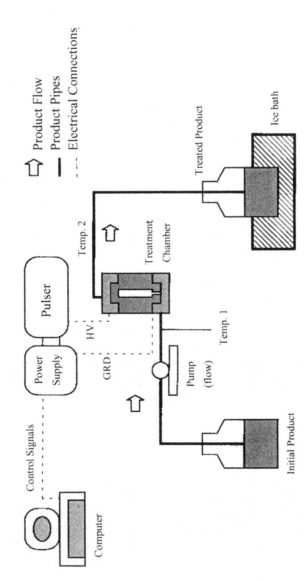

Figure 7.2 Pulser configuration used during the PEF treatments.

Product Flow
Product Pipes
Electrical Connections

Computer

Control Signals

Power Supply

Pulser

HV

GRD

Treatment Chamber

Temp. 2

Temp. 1

Pump (flow)

Initial Product

Treated Product

Ice bath

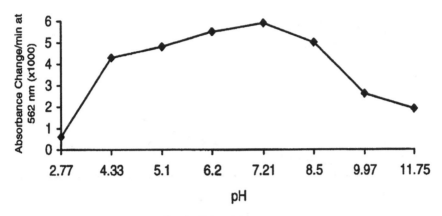

Figure 7.3 Activity of protease from *P. fluorescens* M3/6 at various pH.

The data were then analyzed at a significance level of 5% using the general linear model of SAS© (SAS, 1990).

RESULTS AND DISCUSSION

The partially purified protease from *P. fluorescens* M3/6 cultured in TSB/YE had an optimum pH of 7.2 at 37°C, with casein as the substrate, and a molecular weight in the range of 40 to 50 kDa. This is similar to the reported values by Kohlman et al. (1991). The proteolytic activity of the protease as a function of pH and temperature is summarized in Figures 7.3 and 7.4.

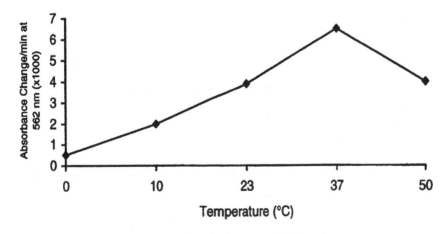

Figure 7.4 Activity of protease from *P. fluorescens* M3/6 at various temperatures.

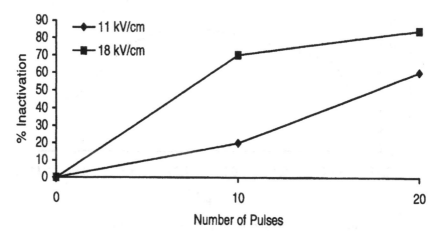

Figure 7.5 Inactivation of protease from *P. fluorescens* M3/6 in TSB/YE at 0.1 Hz.

The inactivation of the protease is a function of the electric field intensity and the number of pulses (Figures 7.5 and 7.6). A reduction of 80% in proteolytic activity of the protease suspended in TSB/YE was observed with a PEF treatment of 20 pulses at 18 kV/cm. A reduction of 60% in proteolytic activity was obtained with 20 pulses of 11 kV/cm (Figure 7.5).

Table 7.3 and Figure 7.6 summarize the inactivation results for the protease when suspended in skim milk and treated with PEF. The PEF inactivation of the protease in skim milk did not follow a similar pattern of the inactivation of the protease in TSB/YE (Figure 7.5).

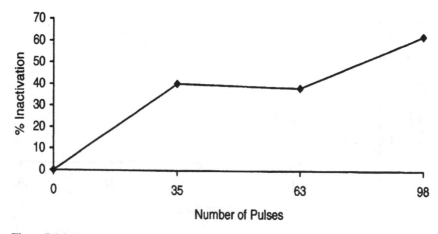

Figure 7.6 Inactivation of protease from *P. fluorescens* M3/6 in skim milk at 15 kV/cm and 2 Hz.

The inactivation of the protease suspended in skim milk was achieved by electric field intensities of 14 and 15 kV/cm and 1 to 2 Hz pulsing rates, with no significant change in the casein susceptibility to proteolysis ($p < 0.05$). Meanwhile, the proteolytic activity and susceptibility of casein to proteolysis were increased when a 25 kV/cm electric field at 0.6 Hz treatment was applied to the skim milk solutions. The proteolysis of the whey proteins was not considered, due to the finding by Bengtsson et al. (1973) that protease from *Pseudomonas* attack casein in the same way as rennet.

The experiments with the casein-Tris buffer demonstrated no significant effect ($p < 0.05$) of PEF on the inactivation of the protease suspended in the casein-Tris buffer (Table 7.5), and there was also no significant change ($p < 0.05$) in the susceptibility of casein to proteolysis when the casein-Tris buffer was exposed to PEF.

The inactivation capability of PEF on the protease from *P. fluorescens* M3/6 was found to be a function of the medium containing the protease. Skim milk and the casein-Tris buffer contained casein, while the TSB/YE mixture was almost free of the substrate. The presence of casein increased the stability of the protease against PEF treatment because of the protection the casein provided to the enzyme as a substrate. Drenth (1981) and Sluyterman (1981) reported an increased stability of proteolytic enzymes when attached to a substrate because of conformational changes induced by charge modifications and hydrophobic interactions in the enzyme-substrate complex.

No significant change in the activity of the protease and no significant increase in susceptibility of the casein to proteolysis in the casein-Tris buffer indicate that the effect of PEF on the protease and casein micelle is a function of chemical species contained in skim milk in addition to casein (e.g., lactose, whey protein).

The limited susceptibility of casein to proteolysis in skim milk treated with 14 or 15 kV/cm at 1 or 2 Hz may account for the inactivation of the protease when exposed to similar PEF conditions while suspended in skim milk. The exposure of skim milk to 25 kV/cm at 0.6 Hz did increase casein susceptibility to proteolysis, which was found to be a better substrate for the protease than untreated skim milk (Table 7.4). The casein in the casein-Tris buffer was an even better substrate for the protease than skim milk, which probably provided a better protection through the enzyme-substrate complex against PEF treatment (Table 7.5).

CONCLUSIONS

The inactivation of an extracellular protease from *Pseudomonas fluorescens* M3/6 by PEF was found to be a function of electric field intensity, number of pulses, and the medium containing the protease. Casein as a substrate was found to protect the protease from being inactivated by PEF treatments.

TABLE 7.4.   PEF Treatments of *P. fluorescens* M3/6 Protease in Skim Milk and Susceptibility to Proteolysis of Proteins in Skim Milk.

	Relative Activity (%)	Relative Susceptibility (%)
Experiment 1: 25 kV/cm		
8 pulses (450 J)	260*	N/T
16 pulses (900 J)	268*	255*
Experiment 2: 14 kV/cm		
16 pulses (298 J)	93.4	N/T
32 pulses (605 J)	62.6*	96.7

*Indicates treatments with a significant difference ($p < 0.05$) from the specific activity of its respective control sample.
N/T: not tested.

The presence of non-casein components such as lactose, whey proteins, and calcium in skim milk are most likely related to the inactivation of the protease and enhanced proteolytic activity in skim milk exposed to a measured electric field of 25 kV/cm. Reduced electric fields (14 or 15 kV/cm) inactivated the protease suspended in the skim milk with no effect on its susceptibility to proteolysis.

ACKNOWLEDGEMENTS

This work was funded by the AASERT Program (U.S. Department of Defense); the Natick Research Development and Engineering Center (U.S. Army); Bonneville Power Administration (U.S. Department of Energy); and Sigma Xi, The Scientific Research Society Grant-in-Aid of Research. Olga Martín-Belloso would also like to express appreciation for the support received from NATO (North Atlantic Treaty Organization) during her sabbatical leave at Washington State University in Pullman, WA.

TABLE 7.5.   PEF Treatments of *P. fluorescens* M3/6 Protease in Casein-Tris Buffer and Susceptibility to Proteolysis of Casein in Casein-Tris Buffer.

	Relative Activity (%)	Relative Susceptibility (%)
Experiment 4: 31.6 kV/cm		
8 pulses (722 J)	109	N/T
16 pulses (1444 J)	104	111
Experiment 5: 23.3 kV/cm		
16 pulses (784 J)	108	N/T
32 pulses (1568 J)	101	139

*Indicates treatments with a significant difference ($p < 0.05$) from the specific activity of its respective control sample.
N/T: not tested.

REFERENCES

Adams, D. M., Barach, J. T., and Speck, M. L. 1975. Heat resistant proteases produced in milk by psychrotrophic bacteria of dairy origin. *J. Dairy Sci.* 58:828–834.

Barach, J. T. and Adams, D. M. 1977. Thermostability at ultrahigh temperatures of thermolysin and protease from a psychrotrophic *Pseudomonas. Biochim. Biophys. Acta.* 485:417–423.

Barach, J. T., Adams, D. M., and Speck, M. L. 1976a. Low temperature inactivation in milk of heat-resistant protease from psychrotrophic bacteria. *J. Dairy Sci.* 59:391–395.

Barach, J. T., Adams, D. M., and Speck, M. L. 1976b. Stabilization of a psychrotrophic *Pseudomonas* protease by calcium against thermal inactivation in milk at ultrahigh temperature. *Appl. Env. Microbiol.* 31:875–879.

Barach, J. T., Adams, D. M., and Speck, M. L. 1978. Mechanism of low temperature inactivation of a heat-resistant bacterial protease in milk. *J. Dairy Sci.* 61:523–528.

Bengtsson, K., Gardhage, L., and Isaksson, B. 1973. Gelation in UHT treated milk, whey and casein solution. The effect of heat resistant proteases. *Mlchwissenschaft.* 28:495–499.

Carlez, A., Rosec, J. P., Richard, N., and Cheftel, J. C. 1993. High pressure inactivation of *Citrobacter freundii, Pseudomonas fluorescens* and *Listeria innocua* in inoculated minced beef muscle. *Lebensm.-Wiss. u-Technol.* 26:357–363.

Castro, A. J. 1994. Pulsed electric field modification of activity and denaturation of alkaline phosphatase. Ph.D. Dissertation. Washington State University, Pullman. WA.

Christen, G. L. and Marshall, R. T. 1984. Thermostability of lipase and protease of *Pseudomonas fluorescens* 27 produced in various broths. *J. Dairy Sci.* 67:1688–1693.

Christen, G. L., Wang, W. C., and Ren, T. J. 1986. Comparison of the heat resistance of bacterial lipases and proteases and the effect on ultra-high temperature milk quality. *J. Dairy Sci.* 69: 2769–2779.

Cousin, M. A. 1982. Presence and activity of psychrotrophic microorganisms in milk and dairy products: A review. *J. Food Prot.* 45:172–207.

Dabbah, R., Moats, W. A., and Mattick, J. I. 1969. Factors affecting resistance to heat and recovery of heat-injured bacteria. *J. Dairy Sci.* 52:608–614.

Drenth, J. 1981. The three dimensional structure of proteolytic enzymes. *Neth. Milk Dairy J.* 35:197–208.

Gilliland, S. E. and Speck, M. L. 1967. Mechanism of the bactericidal action produced by electrohydraulic shock. *Applied Microbiology.* 9:1033–1044.

Hülsheger, H., Potel, J., and Niemann, E. G. 1983. Electric field effects on bacteria and yeast cells. *Radiat. Environ. Biophys.* 22:149–162.

Jacob, H. E., Foster, W., and Berg, H. 1981. Microbial implication of electric field effects. II. Inactivation of yeast cells and repair of their cell envelope. *Z. Allg. Mikrobiol.* 21:225–233.

Kohlman, K. L., Nielsen, S. S., Steenson, L. R., and Ladisch, M. R. 1991. Production of proteases by psychrotrophic microorganisms. *J. Dairy Sci.* 74:3275–3283.

Law, B. A. 1979. Reviews of the progress of dairy science: Enzymes of psychrotrophic bacteria and their effects on milk and milk products. *J. Dairy Res.* 46:573–588.

Law, B. A., Andrew, A. T., and Elisabeth, M. 1977. Gelation of ultra-high-temperature-sterilized milk by proteases from a strain of *Pseudomonas fluorescens* isolated from raw milk. *J. Dairy Res.* 44:145–148.

Malik, A. C. and Swanson, M. 1974. Heat-stable proteases from psychrotrophic bacteria in milk. *J. Dairy Sci.* 57:591–592.

Manji, B. and Kakuda, Y. 1988. The role of protein denaturation, extent of proteolysis, and storage temperature on the mechanism of age gelation in a model system. *J. Dairy Sci.* 71:1455–1463.

Manji, B., Kakuda, Y., and Amott, D. R. 1986. Effect of storage temperature on age gelation of ultra-high temperature milk processed by direct and indirect heating systems. *J. Dairy Sci.* 69:2994–3001.

McKellar, R. C. 1981. Development of off-flavor in ultra-high temperature and pasteurized milk as a function of proteolysis. *J. Dairy Sci.* 64:2138–2145.

Mizuno, A. and Hori, Y. 1988. Destruction of living cells by pulsed high-voltage application. *IEEE Trans. and Ind. Applic.* 24:387–394.

Pothakamury, U. R., Vega-Mercado, H., Zhang, Q., Barbosa-Cánovas, G. V., and Swanson, B. G. 1996. Effect of growth stage and processing temperature on the inactivation of *E. coli* by pulsed electric fields. *J. of Food Protection.* 59(11):1167–1171.

Sale, A. J. H. and Hamilton, W. A. 1967. Effect of high electric fields on microorganisms. 1. Killing of bacteria and yeast. *Biochim. Biophys. Acta.* 48:781–788.

SAS. 1990. *SAS Procedures Guide.* Third edition. Version 6. SAS Institute Inc. Cary, NC.

Sato, M., Tokita, K., Sadakata, M., and Sakai, T. 1988. Sterilization of microorganisms by high-voltage pulsed discharge under water. *Kagaku Hogaku Ronbunshu.* 4:556–557.

Sluyterman, L. A. A. 1981. Chemical aspects of proteolysis and of substrate specificity. *Neth. Milk Dairy J.* 35:209–221.

Zhang, Q., Monsalve-González, A., Barbosa-Cánovas, G. V., and Swanson, B. G. 1994. Inactivation of *E. coli* and *S. cerevisiae* by pulsed electric fields under controlled temperature condition. *Trans. ASAE.* 37:581–587.

CHAPTER 8

Effect of Added Calcium and EDTA on the Inactivation of a Protease from *Pseudomonas fluorescens* M3/6 When Exposed to Pulsed Electric Fields

H. VEGA-MERCADO
J. R. POWERS
G. V. BARBOSA-CÁNOVAS
B. G. SWANSON

ABSTRACT

A N extracellular protease from *Pseudomonas fluorescens* M3/6 was isolated using chromatofocusing and gel exclusion chromatography. It was found to have a molecular weight of 45–50 kDa, an isoelectric point of 8.0, and to be EDTA sensitive. Pulsed electric field (PEF) treatments of 0, 10, and 20 pulses of 700 μs at 6.2 kV/cm and 15°C to 20°C were selected to inactivate the protease suspended in simulated milk ultrafiltrate (SMUF). Thermal inactivation of the protease suspended in SMUF was also evaluated at 55°C for 5, 20, 35, 50, 65, and 80 m. The calcium content in SMUF was adjusted to 0, 10, and 15 mM; the EDTA content to 0, 5, 10, and 20 mM; and the ionic strength adjusted with KCl. It was found that PEF inactivation of the protease was not a function of the concentration of calcium in the media. However, the addition of calcium retarded thermal inactivation of protease, while EDTA enhanced it.

INTRODUCTION

Heat stable enzymes, such as proteases and lipases produced by psychrotrophic microorganisms, alter refrigerated milk and eventually cause spoilage (Law, 1979; Cousin, 1982). Psychrotrophic bacteria such as *Pseudomonas*

121

fluorescens are ubiquitous in nature, and are common contaminants of milk. Bitter flavors and coagulation can occur due to the proteolysis of casein and whey protein by microbial proteases (Malik and Swanson, 1974; Law et al., 1977; Law, 1979; Cousin, 1982; Manji and Kakuda, 1988; McKellar, 1981). The proteases produced by *Pseudomonas fluorescens* are thiol proteinases, with a neutral pH, that are optimal when casein is acting as a substrate. These proteases are ethylenediamine tetra-acetic acid (EDTA) sensitive, with an isoelectric point of pH 8.0 (Barach et al., 1976b; Barach and Adams, 1977). A molecular weight of 45,000 to 50,000 Da has been reported for *Pseudomonas fluorescens* extracellular proteases (Law, 1979).

The high temperature stability of proteases from *Pseudomonas* is related to structural flexibility and the interplay of divalent cations that allow enzyme reactivation after heating (Barach and Adams, 1977). Low temperature (60°C) inactivation requires up to 60 min to achieve 56 to 60% inactivation (Barach et al., 1976a). The heat inactivation of proteolytic enzymes from *Pseudomonas* is explained in terms of the aggregation of proteins and autolysis (Barach et al., 1978).

In general, the active center of an enzyme consists of amino acid residues brought together in the three-dimensional structure of the macromolecule under a delicate balance of different noncovalent forces (e.g., hydrogen bonds, hydrophobic, ionic, van der Waals) (Klibanov, 1983). Enzyme inactivation occurs when the molecule unfolds under certain physical or chemical stress (e.g., heat, pH proteolysis, oxidation) that disassemble the active center (Joly, 1965), heat being the most important inactivating agent (Klibanov, 1983). Tsou (1993) and Zhou et al. (1993) reported that the inactivation of fumarase and creatine kinase by urea or guanidinium chloride (GdmCl) occurs before noticeable conformational changes of the enzyme molecule as a whole. This suggests that enzyme active sites are more susceptible to perturbation by denaturants than the molecule as a whole. Zhang et al. (1993) reported similar results on the thermal inactivation of adenylate kinase.

The inactivation of microorganisms and enzymes using PEF has been extensively studied (Castro, 1994; Pothakamury et al., 1996; Zhang et al., 1994; Sitzmann, 1995; Vega-Mercado, 1996; Vega-Mercado et al., 1995, 1996). The main objective of PEF is to deliver a shelf stable food product with fresh-like taste, nutrient content, and good appearance. It has been reported that the inactivation of microorganisms exposed to PEF is mainly related to changes in cell membrane permeability and electromechanical instability (Jacob et al., 1981; Coster and Zimmermann, 1975).

Gilliland and Speck (1967) reported the inactivation of trypsin, lactic dehydrogenase, and protease from *B. subtilis* using an electric field of 31.5 kV/cm, and Castro (1994) reported the inactivation of alkaline phosphatase in SMUF by applying a PEF of 20 kV/cm. The enzyme inactivation mechanism proposed by Gilliland and Speck (1967) was an oxidative reaction induced by the electric field as a function of treatment time. Castro (1994) proposed a conformational change in which hydrophobic groups within the enzyme are realigned and combined to cause aggregation and, thus, inactivation of the enzyme.

This study was designed to examine the effect of calcium and EDTA on an extracellular protease from *Pseudomonas fluorescens* M3/6 when exposed to PEF. The variables under consideration for this study include the number of pulses and composition of the medium where the enzyme was exposed to a constant electric field of 6.2 kV/cm in pulses of 700 μs.

MATERIALS AND METHODS

GROWTH OF *PSEUDOMONAS FLUORESCENS* M3/6, PROTEASE COLLECTION, AND PARTIAL CHARACTERIZATION

The *Pseudomonas fluorescens* M3/6 was obtained from Dr. S. S. Nielsen, Food Science Department, Purdue University. Tryptic soy broth and agar (TSB, TSA) enriched with yeast extract (YE, 0.6% w/v) were used to grow the cells in a 28-day cycle to preserve their proteolytic activity.

Sterile TSB/YE (in batches of 300 mL) was inoculated with the *P. fluorescens* M3/6 and incubated at 25°C for 24 hours before it was transferred to refrigeration temperature of 4°C to 6°C for 18 days. The cells were then removed by centrifugation at 4700 × g for 5 min, and ultrafiltration using a 0.45 μm sterile vacuum filter. The filtered solution was tested for proteolytic activity using the Pierce QuantiCleave Protease Assay Kit (Pierce, Rockford, IL).

CONCENTRATION AND CHROMATOGRAPHIC PURIFICATION OF THE PROTEASE

Concentration of the TSB/YE-Protease Mixture

The molecular weight of the chemical species contained in the sterile TSB/YE-protease mixture with pH 7.0 was narrowed by ultrafiltration using 50 kDa and 30 kDa membrane filters (Microsep[TM], Filtron Technology Corp., Northborough, MA), whose selection was based on the reported molecular weight for the protease from *P. fluorescens* M3/6 (Kohlman et al., 1991). The TSB/YE-protease mixture was ultrafiltered using a 50 kDa filter. A sample of 3 mL was placed in the upper reservoir of the filter and centrifuged at 4700 × g for 60 min at 0°C. The filtrate from the 50 kDa filter was then collected and transferred to the reservoir of a 30 kDa filter. The filtration-concentration procedure was carried out under the same conditions as those used for the 50 kDa. The 30 kDa material was collected, filtered through a 0.45 μm sterile filter and stored at

−20°C in sterile test tubes. The procedure was repeated until enough material of partially purified mixture was obtained.

Purification of TSB/YE-Protease Mixture

A pH gradient was formed by equilibrating a polybuffer exchange column (PBE, Pharmacia Biotech AB, Uppsala, Sweden) with 0.025 M ethanolamine-acetic acid, and eluted with Polybuffer96 (Pharmacia Biotech AB, Uppsala, Sweden) to separate the proteins in the TSB/YE-enzyme mixture. The PBE column, with a bed volume of 30 mL, was attached to the WATERS 650E (Millipore Corp., Milford, MA) protein purification system equipped with a UV detector at 280 nm wavelength. The column was equilibrated to pH 9.4 using 0.025 M ethanolamine-CH_3COOH at 2 mL/min. Once equilibrated, 4 mL of Polybuffer96 (pH 7.0) was applied to the column, followed by an injection of 5 mL of TSB/YE-protease concentrate. The sample was eluted using the Polybuffer96 at a flow rate of 1.0 mL/min. The pH (0.5 mL) of the eluted sample's fractions was monitored using an Orion 420A pHmeter (Orion Research Inc., Boston, MA) equipped with a pencil-thin gel-filled combination electrode (Accumet, Fisher Scientific, Pittsburgh, PA). The elution continued until a pH of 7.0 was obtained for the collected fractions. The PBE column was regenerated using a mixture of 1.0 M NaCl and 0.025 M ethanolamine-CH_3COOH with a pH of 9.4 at a flow rate of 2 mL/min to remove any bound protein not eluted during the Polybuffer96 elution. The column was then re-equilibrated using the 0.025 M ethanolamine (pH 9.4) to remove the NaCl. Collected fractions were assayed for activity and protein content. Active fractions were pooled and concentrated to 1.0 mL using a 10 kDA filter.

A Waters Protein Pak Glass 300 SW column (Millipore Corp., Milford, MA), attached to a WATERS 650E (Millipore Corp., Milford, MA) protein purification system equipped with a UV detector at 280 nm wavelength, was used for size classification of the active fractions obtained from the chromatofocusing procedure. The column was equilibrated using 0.05 M Tris-HCl buffer with a pH of 7.5 and 0.05 M NaCl. Gel filtration molecular weight markers (SIGMA Chemical Company, St. Louis, MO) were used as standards for the size classification procedure. Molecular weight markers and Polybuffer96 protein concentrate were applied to the gel exclusion column and eluted at 0.5 mL/min. Protein fractions of 1.0 mL were collected and analyzed for proteolytic activity and protein content.

PHYSICOCHEMICAL EFFECT OF HEAT AND PEF ON THE PROTEASE

A Hydrophobic Interaction Column (HIC) (BIO-RAD, Richmond, CA), attached to a WATERS 650E (Millipore Corp., Milford, MA) protein purification system equipped with a UV detector at 280 nm wavelength, was used to evaluate the effect of heat and PEF treatments on the protease from *Pseudomonas fluorescens* M3/6. The HIC column was equilibrated using 0.05 M Tris-HCl buffer with a pH of 7.5 with 0.05 M NaCl. Samples were applied to the HIC column at 0.5 mL/min and the elution time and peak absorbance at 280 nm were recorded for each case.

SDS-PAGE ELECTROPHORESIS

The molecular weight distribution of the proteins in the TSB/YE-protease mixture, Polybuffer96 fractions, and gel exclusion fractions was determined using 4% to 20% Gradient Mini-Protean II Ready Gels (BIO-RAD, Richmond, CA). The electrophoresis was performed using a Mini-PROTEAN II cell (BIO-RAD, Richmond, CA) at a constant voltage (200 V d.c.) for 30 min. Silver staining of the gels was carried out as recommended by the manufacturer (BIO-RAD, Richmond, CA). Broad molecular weight standards (BIO-RAD, Richmond, CA) containing myosin, β-galactosidase, phosphorylase b, bovine serum albumin, ovalbumin, carbonic anhydrase, soybean trypsin inhibitor, lysozyme, and aprotinin were used for the molecular weight determination.

Protein samples for electrophoresis were precipitated using 25% trichloroacetic acid (TCA). Three hundred μL of protein samples were mixed with a similar volume of TCA, and centrifuged at 6000 × g for 15 min. The supernatant was carefully removed, 50 μL of sample buffer (containing bromophenol blue and beta-mercaptoethanol) was added into the vial, and pH was adjusted by adding 2 μL of saturated Tris-base, vortexing vigorously, and heating in boiling water for 5 min. The electrophoresis gel was loaded with 25 μL of a sample buffer containing the proteins.

SIMULATED MILK ULTRAFILTRATE FOR THE INACTIVATION OF THE PROTEASE

Protease samples (150 μg) were suspended in 3 mL of SMUF with a pH of 7.0 for PEF and heating experiments. The SMUF stock solution consisted of 138.7 mM lactose, 11.6 mM KH_2PO_4, 3.3 mM tripotassium citrate, 6.0 mM trisodium citrate, 1.8 mM magnesium citrate, and 2.2 mM potassium carbonate. Three different concentrations of calcium and four different concentrations of EDTA were used to evaluate the inhibitory effect of PEF and heat on the protease.

The calcium concentration was adjusted to 0, 10, and 15 mM by adding calcium chloride to the SMUF stock solution. The ionic strength of the SMUF was adjusted to 168 mM using potassium chloride (Table 8.1). Four levels of EDTA (0, 5, 10, 20 mM) were selected to evaluate the inhibitory effect of the

TABLE 8.1. Calcium and EDTA Content of SMUF.

Formulation	Ca^{++} (mM)	KCl (mM)	EDTA (mM)
1	0	44	0
2	0	44	5
3	0	44	10
4	0	44	20
5	10	19	0
6	15	6.7	0

chelating agent on the protease. EDTA was added to the stock solution and the ionic strength was adjusted with KCl.

PEF AND HEAT INACTIVATION EXPERIMENTS

PEF treatments using a constant electric field of 6.2 kV/cm and 0, 10, and 20 pulses of 700 Vs were applied using a bench top electroporator (GeneZapper, IBI-Kodak, Rochester, NY). One mL of SMUF containing the protease (50 μg/mL) was deposited into a 0.4 cm gap cuvette. The pulsing rate was approximately 0.1 Hz, and the temperature of the cuvette was 15°C to 20°C. The cuvette was immersed in cold water (10°C) every 5 pulses and the PEF treatments using all six SMUF formulations were done in triplicate.

Heating experiments using a constant temperature of 55°C for 0, 5, 20, 35, 50, 65, and 80 min were selected to evaluate the thermal inactivation of the protease suspended in SMUF. Three mL of the SMUF containing the protease (50 μg/mL) were immersed in a water bath at 55°C. The come-up time for the SMUF to reach 55°C was 5 min. Aliquots of 500 μL were taken, cooled in ice, and analyzed for their proteolytic activity. SMUF formulations containing calcium chloride and potassium chloride were selected for duplicate heating experiments.

ENZYMATIC AND PROTEIN ASSAYS

Bicinchoninic acid reagent (BCA*) was prepared by mixing 50 parts bicinchoninic acid one part copper sulfate solution provided by the manufacturer (Pierce, Rockford, IL). Once mixed, the BCA* was dispensed into glass vials (2 mL/vial), covered, and stored at room temperature until further use.

Proteolytic Assay

An aliquot of the enzyme (500 μL) was mixed with a 1% casein solution (2000 μL) provided with the QuantiCleave Protease Assay Kit (Pierce,

Rockford, IL). The mixture was incubated at 37°C. A sample of 1.0 mL was removed at 0 and 20 hr for analysis. Trichloroacetic acid (TCA, 60 μL, 6.12 molal) was added to precipitate the nondigested casein. The mixture was filtered using a 0.45 μm filter to remove the insoluble fraction. A 100 μL aliquot from each filtrate was added to 2 mL BCA* and incubated for 30 min at 37°C. The absorbance was measured at 562 nm using an HP 8452A diode array spectrophotomer (Hewlett Packard, Waldbronn, Germany) (Figure 8.1).

A standard curve using bovine serum albumin (BSA) was used as a reference. The proteolysis rate and specific activity (sp. ac.) were determined as follows:

$$\text{rate} = \frac{\text{net change absorbance}}{\text{min}}$$

$$\text{sp. ac.} = \frac{\text{net change absorbance}}{\mu\text{g protein} - \text{min}}$$

The inactivation was evaluated as the percent of reduction in the specific activity after the PEF treatments.

Total Protein Assay

The BCA* reagent was used to estimate the amount of enzyme in the solution. Aliquots of the enzyme preparation (100 μL) were added to 2 mL of BCA* and incubated for 30 min at 37°C. The amount of protein was determined from a BSA standard curve.

RESULTS AND DISCUSSION

PURIFICATION OF THE PROTEASE

The protease from *Pseudomonas fluorescens* M3/6 was isolated from a tryptic soy broth enriched with a yeast extract. The amount of enzyme in the TSB/YE mixture represented approximately 0.2% of the total protein content for the mixture. The isoelectric point of the protease was estimated at 8.0 using the chromatofocusing technique. Figure 8.2 illustrates the elution, pH and activity profiles for the chromatofocusing separation of the TSB/YE-protease mixture. A pool of the active protein fractions from the chromatofocusing procedure, once mixed and concentrated, was loaded onto a gel exclusion column for molecular weight classification as illustrated in Figure 8.3. Figure 8.4 illustrates the distribution (by SDS-PAGE) of proteins for each step of the

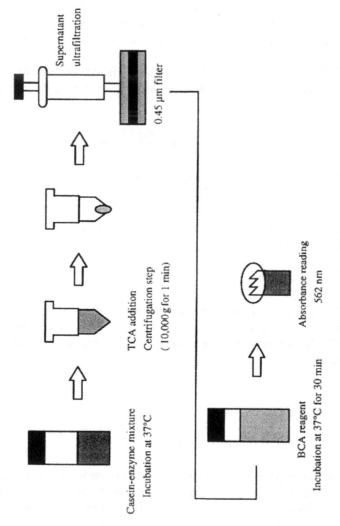

Casein-enzyme mixture
Incubation at 37°C

TCA addition
Centrifugation step
(10,000 g for 1 min)

Supernatant
ultrafiltration

0.45 µm filter

BCA reagent
Incubation at 37°C for 30 min

Absorbance reading
562 nm

Figure 8.1 Enzymatic activity assay procedure.

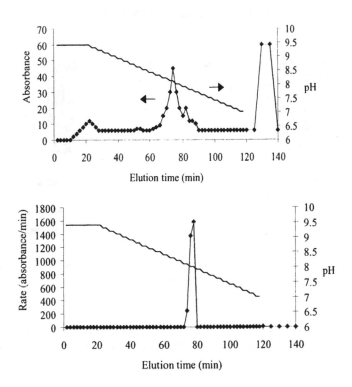

Figure 8.2 Elution, pH, and activity profiles for chromatofocusing of TSB/YE protease mixtures: (a) protein profile, and (b) activity profile.

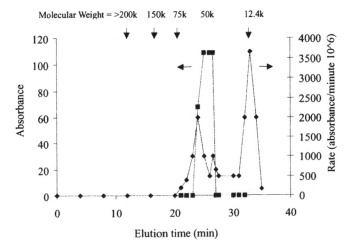

Figure 8.3 Elution and activity profile for gel exclusion chromatography of partially purified TBS/YE protease mixture from *Pseudomonas fluorescens* M3/6.

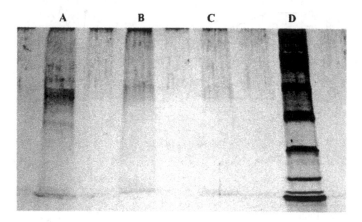

Figure 8.4 Silver staining of electrophoresis gel: (a) TSB/YE-protease mixture, (b) Polybuffer96 pool, (c) Gel exclusion pool; (d) reference molecular weight.

purification procedure. Table 8.2 illustrates the yield for each step during the purification of the protease. A sixfold increase in specific activity was obtained overall.

INHIBITORY EFFECT OF PEF, HEAT, AND EDTA ON THE PROTEASE

Inactivation of the protease from *Pseudomonas fluorescens* M3/6 when exposed to PEF did not depend on the presence of calcium in the media containing the protease, as illustrated in Figure 8.5. The inactivation was the same for all three solutions containing 0, 10, or 15 mM calcium. The proteolytic activity of the protease was reduced 30% after exposure to 20 pulses of 700 μs at 6.2 kV/cm and 15–20°C.

In contrast to PEF, thermal inactivation of the protease suspended in the SMUF did vary with calcium content. Samples containing either 10 or 15 mM

TABLE 8.2. Proteolytic Activity of Protein Samples.

Sample	Sp. Act. (abs/min μ 10^{-6})	Protein (μg/mL)	Volume (mL)	Total Activity (abs/min)	Yield (%)
TSB/YE-protease	1.84	3000	300	1.65	100
Polybuffer96	4.93	500	8	0.02	1.2
Gel exclusion	12.68	500	1.5	0.01	0.7

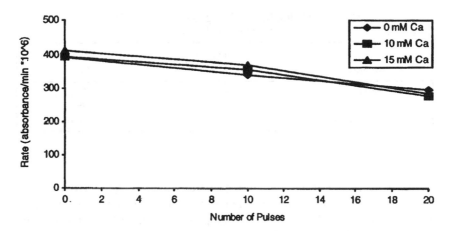

Figure 8.5 PEF inactivation of protease from *Pseudomonas fluorescens* M3/6 at 6.2 kV/cm.

calcium retained 71% of their original activity compared to 12% retention in the samples without calcium after 5 min of heating at 55°C, followed by a steady decrease in activity as a function of the heating time (Figure 8.6).

The analysis using the hydrophobic interaction column (HIC) of PEF (20 pulses, 15 mm Ca^{++}) and heat treated (5 min, 15 mM Ca^{++}) samples showed differences in the retention time and peak high of the eluted protein when compared to the nontreated samples (Table 8.3). The increase in the retention time indicated unfolding.

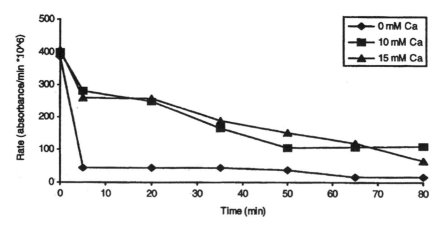

Figure 8.6 Thermal inactivation of protease from *Pseudomonas fluorescens* M3/6 at 55°C.

TABLE 8.3. Hydrophobic Changes Induced by PEF and
Thermal Treatments.

Sample	Retention Time (min)	Peak High (mm)
Control	6.01	22.9
20 Pulses	5.96	25.4
Heat treated	5.93	20.6

EDTA had a significant inhibitory effect on the proteolytic activity of the protease (Figure 8.7). This result is similar to that reported data for the protease from *P. fluorescens*. PEF treatment of samples containing EDTA enhanced the inactivation of the protease (Figure 8.8).

CONCLUSIONS

An extracellular protease from *Pseudomonas fluorescens* M3/6 was isolated using chromatofocusing and gel exclusion chromatography. It was found to have a molecular weight of 45–50 kDa, an isoelectric point at pH 8.0, and be EDTA sensitive. PEF inactivation of the protease was not a function of the calcium added, while thermal inactivation varied significantly when calcium was not present in SMUF. The chelating action of the EDTA enhanced the inactivation of the protease when treated with PEF. Preliminary results from

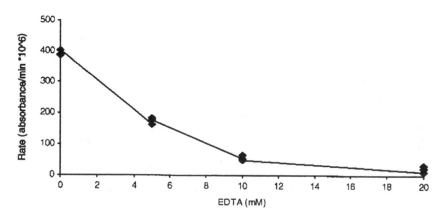

Figure 8.7 Inhibitory effect of EDTA on a protease from *Pseudomonas fluorescens* M3/6.

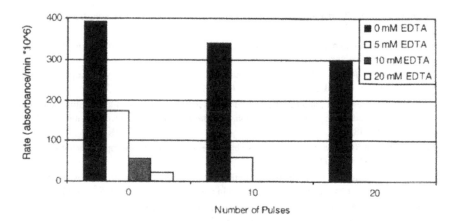

Figure 8.8 PEF inactivation of a protease from *Pseudomonas fluorescens* M3/6 in SMUF with EDTA.

the HPLC using the hydrophobic interaction column indicated changes in the conformation of the protein after PEF treatments.

REFERENCES

Barach, J. T. and Adams, D. M. 1977. Thermostability at ultrahigh temperatures of thermolysin and protease from a psychrotrophic *Pseudomonas. Biochim. Biophys. Acta.* 485:417–423.

Barach, J. T., Adams, D. M., and Speck, M. L. 1976a. Low temperature inactivation in milk of heat-resistant protease from psychrotrophic bacteria. *J. Dairy Sci.* 59:391–395.

Barach, J. T., Adams, D. M., and Speck, M. L. 1976b. Stabilization of a psychrotrophic *Pseudomonas* protease by calcium against thermal inactivation in milk at ultrahigh temperature. *Appl. Env. Microbiol.* 31:875–879.

Barach, J. T., Adams, D. M., and Speck, M. L. 1978. Mechanism of low temperature inactivation of a heat-resistant bacterial protease in milk. *J. Dairy Sci.* 61:523–528.

Castro, A. J. 1994. Pulsed electric field modification of activity and denaturation of alkaline phosphatase. Ph.D. dissertation, Washington State University, Pullman. WA.

Coster, H. G. L. and Zimmermann, U. 1975. The mechanisms of electrical breakdown in the membrane of *Valonia utricularis. J. Membrane Biol.* 22:73–90.

Cousin, M. A. 1982. Presence and activity of psychrotrophic microorganisms in milk and dairy products: A review. *J. Food Prot.* 45:172–207.

Gilliland, S. E. and Speck, M. L. 1967. Mechanism of the bactericidal action produced by electrohydraulic shock. *Applied Microbiol.* 9:1033–1044.

Jacob, H. E., Foster, W., and Berg, H. 1981. Microbial implication of electric field effects. II. Inactivation of yeast cells and repair of their cell envelope. *Z. Allg. Mikrobiol.* 21:225–233.

Joly, M. 1965. A Physicochemical Approach to the Denaturation of Proteins. Academic Press, NY.

Klibanov, A. M. 1983. Stabilization of enzymes against thermal inactivation. *Adv. Appl. Microbiol.* 29:1–28.

Kohlman, K. L., Nielsen, S. S., Steenson, L. R., and Ladisch, M. R. 1991. Production of proteases by psychrotrophic microorganisms. *J. Dairy Sci.* 74:3275–3283.

Law, B. A. 1979. Reviews of the progress of dairy science: Enzymes of psychrotrophic bacteria and their effects on milk and milk products. *J. Dairy Res.* 46:573–588.

Law, B. A., Andrew, A. T., and Elisabeth, M. 1977. Gelation of ultra-high-temperature-sterilized milk by proteases from a strain of *Pseudomonas fluorescens* isolated from raw milk. *J. Dairy Res.* 44:145–148.

Malik, A. C. and Swanson, M. 1974. Heat-stable proteases from psychrotrophic bacteria in milk. *J. Dairy Sci.* 57:591–592.

Manji, B. and Kakuda, Y. 1988. The role of protein denaturation, extent of proteolysis, and storage temperature on the mechanism of age gelation in a model system. *J. Dairy Sci.* 71:1455–1463.

McKellar, R. C. 1981. Development of off-flavor in ultra-high temperature and pasteurized milk as a function of proteolysis. *J. Dairy Sci.* 64:2138–2145.

Pothakamury, U. R., Vega-Mercado, H., Zhang, Q., Barbosa-Cánovas, G. V., and Swanson, B. G. 1996. Effect of growth stage and temperature on the inactivation of *E. coli* by pulsed electric fields. *J. of Food Prot.* 59(11):1167–1171.

Sitzmann, W. 1995. High voltage pulse techniques for food preservation. In: *New Methods of Food Preservation* (G. W. Gould, ed.), Blackie Academic & Professional, London. pp. 236–252.

Tsou, C. L. 1993. Conformational flexibility of enzyme active sites. *Sci.* 262:380–381.

Vega-Mercado, H. 1996. Inactivation of proteolytic enzymes and selected microorganisms in foods using pulsed electric fields. Ph.D. dissertation, Washington State University, Biological Systems Engineering Department, Pullman, WA.

Vega-Mercado, H., Powers, J. R., Barbosa-Cánovas, G. V., and Swanson, B. G. 1995. Plasmin inactivation with pulsed electric fields. *J. Food Sci.* 60:1143–1146.

Vega-Mercado, H., Pothakamury U. R., Chang F. J., Barbosa-Cánovas G. V., and Swanson, B. G. 1996. Inactivation of *E. coli* by combining pH, ionic strength and pulsed electric field hurdles. *Food Res. Int.* 29(2):117–121.

Zhang, Y. L., Zhou, J. M., and Tsou, C. L. 1993. Inactivation precedes conformation change during thermal denaturation of adenylate kinase. *Biochimica et Biophysica Acta.* 1164:61–67.

Zhang, Q., Monsalve-González, A., Barbosa-Cánovas, G. V., Swanson, B. G. 1994. Inactivation of *E. coli* and *S. cerevisiae* by pulsed electric fields under controlled temperature conditions. *Trans. ASAE.* 37:581–587.

Zhou, H. M., Zhang, X. H., Yin, Y., and Tsou, C. L. 1993. Conformational changes at the active site of creatine kinase at low concentrations of guanidinium chloride. *Biochem. J.* 291:103–107.

CHAPTER 9

Nonthermal Inactivation of Endoproteases by Pulsed Electric Field Technology

L. A. PALOMEQUE
M. M. GÓNGORA-NIETO
A. S. BERMÚDEZ
G. V. BARBOSA-CÁNOVAS
B. G. SWANSON

ABSTRACT

PULSED electric field (PEF) treatment was applied to inactivate the proteolytic activity of two commercial endoproteases and to stop the hydrolysis process on faba bean protein concentrates (PC). The enzymatic inactivation capabilities of PEF, as well as the solubility and water absorption capacity of protein hydrolysates stabilized by PEF, were compared to those obtained by two traditional methods: (a) thermal treatment at 92°C for 5 min, and (b) acid conditions lowering the pH to 4.

Faba bean protein hydrolysate suspensions containing 50 μg/mL of an enzyme mixture (Protamex™:Alcalase™, 1:1) were treated with PEF in a continuous system (coaxial treatment chamber) with 1–181 pulses of 2–4 μs and electric fields from 42 to 78 kV/cm. Processing temperature was kept below 32°C. No residual activity of the enzymes was detected after 31 pulses at 78 kV/cm, or at 24, 48, or 72 hours after storage at 4°C. A residual activity of 60% was measurable after the thermal and acidification treatments, therefore they were not as effective as PEF, in inactivating the proteases or stopping the hydrolysis of the PC. Furthermore, the protein solubility index (PSI) and the water absorption capacity (WAC) were higher in those solutions of protein hydrolysates treated with PEF, compared to thermal and acid treatments.

INTRODUCTION

Production of high protein foods from non-conventional sources is of utmost relevance in developing countries where elevated degrees of malnutrition are present. Meanwhile, in developed countries, consumers with a special need for protein supplementation, such as the elderly, athletes, or those on low-calorie diets, are demanding ready-to-eat, low-fat, and high-protein beverages (Frokjaer, 1994). In addition, the use of hydrolysate-based products in special medical diets for the treatment of pancreatitits, short bowel syndrome, Crohn's disease, and food allergies is increasing (Schmidl et al., 1994). Seeds of *Vicia faba* have been found to be an excellent source of high protein concentrates for nutritional supplements (Montilla, 1994; Cantoral et al., 1995) that can be used in the preparation of baby food and special diets for those with enteric nutrition problems (Camacho et al., 1998).

The success of using any seed protein as a food ingredient depends not only on its essential amino acid content, but also on its functional properties. Solubility, fat, and water retention are some of the most important functional properties of protein concentrates (PC). High solubility is needed to increase PC digestibility and to facilitate their incorporation into food supplements. The solubility of faba bean proteins can be increased with enzymatic hydrolysis using endoproteases (Palomeque and Bermúdez, 1997; Baquero and Bermúdez, 1997; Pilosof, 1995; Mahmoud, 1994; Adler-Nissen, 1976). However, it is necessary to stop the hydrolysis process in order to obtain peptides of appropriate sizes that won't produce a dehydration of the small intestine after consumption of the protein hydrolysate-product, as well as to control the development of bitterness due the excessive proteolytic cleavage of proteins (Pedersen, 1994). The inactivation of enzymes using traditional methods (acidification/pH reduction or heat treatment) has some disadvantages; the hydrolyzed protein concentrates (HPC) may present:

- high salt concentration
- low solubility
- changes in organoleptic characteristics
- partial starch gelation

The nonthermal enzyme-inactivating capabilities of pulsed electric fields (PEF) (Ho et al., 1995; Vega-Mercado et al., 1995; Qin et al., 1995) make this treatment an attractive alternative for producing high quality HPC. Pulsed electric field processing is an emerging nonthermal method of food preservation in which a high electric field ($\sim$40 kV/cm) is applied in the form of short pulses ($\sim$1–5 μs) to a fluid food confined in or flowing through a pair of high voltage electrodes. In PEF, the applied electric field and total number of pulses (or total treatment time) are the two main variables responsible for microbial and enzyme inactivation. Pulsed electric field treatments have been demonstrated to process foodlike products in a short period of time, yet achieve the inactivation of spoilage and pathogenic microorganisms, as well as the enzymes responsible for undesirable reactions in foods. In general, it is expected that a minimally processed PEF product will retain its fresh physical, chemical, and nutritional characteristics (Barbosa-Cánovas et al., 1999). Continuous PEF treatment systems have been used to pasteurize foods, such as milk and dairy products, a variety of fruit juices, beaten eggs, and cream soups. In such studies, no significant changes have been detected in either chemical or physical parameters; furthermore, sensory panels have found no significant differences between the PEF treated and non-PEF treated products (Barbosa-Cánovas et al., 1998, 1999; Vega-Mercado et al., 1995; Zhang-Ying and Yan, 1993).

The aim of this study is to evaluate the effect of PEF treatments over an endoprotease complex activity (Protamex™ and Alcalase™), as well as the final HPC solubility and wettability (WAC), in comparison with traditional methods (acidification and heat treatment) of enzymatic inactivation.

MATERIALS AND METHODS

Dry faba bean (*Vicia faba*) seeds were obtained from a local store (convenience store, Bogotá, Colombia). Alcalase 2.4L™, a food-grade serine bacterial endopeptidase with specific activity of 2.4 Anson units per gram (AU/g), and Protamex™, a *Bacillus* protease complex with specific activity of 1.5 AU/g, were donated by NOVO-NORDISK (Enzyme Process Division, Novo Allé, DK-2880 Bagsvaerd, Denmark).

PREPARATION OF HIGH PROTEIN CONCENTRATES (PC)

Seeds of faba beans were milled and sieved to obtain faba bean flour. The powdered faba was suspended in distilled water with a flour/water ratio of 1:6 (W/V) at 40°C. The pH was adjusted to 8.0 by means of a 5 N solution of NaOH to obtain an acceptable nitrogen extraction (Mordecay and Bermúdez, 1994; McCurdy and Knipfel, 1990). The suspension was blended in a semi-industrial blender for 10 min. The supernatant containing the proteins, carbohydrates, and mineral salts was collected, and the solid residual was discarded. The pH of the collected supernatant was adjusted to 5.5 with 5 N HCl to precipitate the proteins that were decanted by gravity (Palomeque and Bermúdez, 1995). The precipitate was dehydrated in a drum/roller dehydrator (REEVES, 40 psi, 140°C, contact time 20 to 30 s) (Palomeque and Bermúdez, 1995).

The protein content of dehydrated faba bean flour and PC was determined by the Kjeldahl method (AOAC, 1990). The faba bean flour contained 25.78% protein (dry basis). The percentage of protein in the faba bean PC was 45%.

TABLE 9.1. Amino Acid Balance (mg in 100 g) of Faba Bean Products.

Amino Acid	In Flour (mg of Dry Matter)	In Protein Concentrate (mg of Dry Matter)
Asparagine	32.88	76
Threonine	12.86	25.61
Serine	16.4	39.31
Glutamin	54.88	125.0
Proline	10.13	26.26
Glycine	12.99	27.7
Alanine	12.74	28.88
Valine	15.11	35.03
Cysteine	1.21	2.07
Methionine	0.44	1.04
Isoleucine	13.24	31.35
Leucine	24.01	56.8
Tyrosine	10.14	24.81
Phenylalanine	13.53	33.04
Lysine	21.78	46.05
Histidine	8.22	18.62
Arginine	28.57	60.97

The amino acid balance of the faba bean flour and PC are reported in Table 9.1.

PROTEIN CONCENTRATE HYDROLYSIS

It has been demonstrated that enzymatic proteolysis with Protamex™ and/or Alcalase™ is useful in increasing the solubility of faba bean protein concentrates (Baquero and Bermúdez, 1997; Mosquera et al., 1996; Niño et al., 1996; Palomeque and Bermúdez, 1995, 1997). The hydrolysis conditions used in the present study were selected based on results obtained by previous reports (Palomeque and Bermúdez, 1995, 1997). The pH of PC suspensions (0.5% protein/water) was adjusted to 7.0, followed by the addition of Protamex™ and Alcalase™ (enzyme/protein 0.5% respectively). The hydrolysis process was conducted at 60°C for 15 min. Immediately after this, the process was stopped by either acidification (adding HCl to pH 4.0), heat (92°C for 5 min), or PEF treatments. The stabilized dispersions were freeze-dried.

PEF ENZYME INACTIVATION

A pilot plant size pulse voltage generator, manufactured by Physics International (San Leandro, CA), was used to conduct the PEF treatments. A high voltage power source, set to 35 kV to 40 kV, charged a 5 μF capacitor, and a spark gap switch was used to deliver exponential decaying pulses to a continuous treatment chamber consisting of coaxial stainless steel electrodes. The duration of the pulse was evaluated as the decay time constant of the system, as defined in Equation (1):

$$\tau = CR \tag{1}$$

where C is the system capacitance and R the load resistance. The pulsing rate was set at 3 Hz, and the average electric field applied to the samples was evaluated with Equation (2):

$$E_{avg} = \left(\frac{2}{R_{hv} R_{lv}} \right) \left(\frac{V_0}{\ln (R_{lv}/R_{hv})} \right) \tag{2}$$

where R_{hv} and R_{lv} are the radii of the high and low voltage electrodes, respectively, and V_0 is the delivered voltage to the PEF chamber determined with an oscilloscope (Hewlett Packard 54520A, Colorado Springs, CO). The radius of the high voltage electrode was adjusted to obtain 0.7 and 0.45 cm gaps. The temperature of the treated samples was observed at the exit of the treatment chamber using a digital thermometer (John Fluke Mfg. Co., Everett, WA). Protein dispersions (0.5% protein/water) that contained the enzymes under study (enzyme/protein 0.5% or 25 µg/mL respectively) were subjected to continuous recirculation PEF treatments using a total volume in the system of 1.5 L. The flow rate of 500 mL/min was controlled by a peristaltic pump (Masterflex Model 7564-00, Cole Parmer Instrument Co., Chicago, IL). Samples were collected at 0, 7, 16, 23, 31, 47, 62, 78, 93, and 109 pulses, when the 0.45 cm gap was used, and at 0, 11, 26, 39, 52, 78, 104, 130, 155, and 181 pulses when the 0.71 cm gap was used. Experiments were conducted in duplicate.

PHYSICOCHEMICAL PROPERTIES ANALYSIS

Proteolytic Activity

The proteolytic activity of a Protamex™ and Alcalase™ mixture was determined using a Quanticleave™ Protease Assay Kit (PIERCE, Rockford, IL). The assay was based on the use of succinylated casein in conjunction with TNBSA (trinitrobenzensulfonic acid). The proteases (Protamex™ and Alcalase™) in the sample (HPC) act on the succinylated casein, supplied in the kit, to randomly cleave bonds and expose primary amines. These amines react with the TNBSA to produce an orange-yellow color that was quantified using a spectrophotometer; A_{450} absorbance data were collected with a Hewlett Packard Spectrophotometer 8452A, HP Diode Array. The lower the A_{450} absorbance readings, the lower the concentration and/or activity of protease mixture in the HPC after the hydrolysis was stopped, independently, by each of the three different treatments (acidification, heat, or PEF).

Protein Solubility

The solubility of the protein concentrates was determined within a pH range of 3.0 to 8.0. Suspensions of PC and HPC (5 mg protein/mL) were prepared using a 0.1 N NaCl solution. The pH was adjusted by adding either 0.1 N HCl or 0.1 N NaOH. The suspension was shaken for 30 min at room temperature and then centrifuged (30 min, 6000 rpm). The solubility was expressed as the protein content in the supernatant (PSI-protein solubility index; Morr et al., 1985), which was evaluated by the Kjeldahl method. Each evaluation was conducted in duplicate, and the average was reported.

Water Absorption Capacity (Wettability)

The Baumann apparatus permits the measurement of spontaneous water uptake by food ingredients. Under conditions where total dissolution is prevented by the presence of various intermolecular forces, water uptake can be observed until equilibrium is reached, since at this point, the sorbent is chemically stable in the presence of the solvent. This method has been used mainly to determine the total amount of water taken up by a sample at equilibrium (Elizalde et al., 1996; De Kanterewicz et al., 1989; Pilosof et al., 1985). The measurement of water absorption capacity (wettability: WAC) of the samples was carried out following the system described by Elizalde et al. (1996). Maximum spontaneous uptake of water by the PC and HPC in the form of dry powders was determined on 50–100 mg samples at 20°C and relative humidity of 75%.

RESULTS AND DISCUSSION

ENZYMATIC ACTIVITY

The quanticlave protease assay kit was used to construct a control curve with 100% enzymatic activity of protease:alcalase mixtures with different concentrations, as shown in Figure 9.1. The curve in Figure 9.1 has a correlation coefficient of 0.9502, thus indicating that the method was appropriated to follow the enzymatic inactivation after treatment (acidification, heat, or PEF) of HPC. However, the correlation of Figure 9.1 was not used to evaluate the residual enzymatic activities in the HPC after the inactivation treatments (acidification, heat, PEF), since the presence in the suspensions of salts, proteins, carbohydrates,

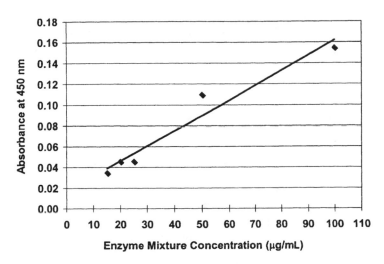

Figure 9.1 Control curve for 100% enzymatic activity.

and additional components interfered in the absorbance readings. Instead, a relative residual activity was evaluated, considering the difference between the A_{450} absorbance readings of the samples before and after the inactivation treatments.

The HPC treated with PEF had an acceptable residual enzymatic activity (30.78%), considering that the percentage of residual enzymatic activity achieved using the traditional inactivation methods was more than 10% higher (acidification: 40.31%, heat: 42.49%). Figure 9.2 shows the enzyme activity of faba HPC as a function of supplied electric field intensity and pulse numbers. The enzyme mixture activity decreased when the applied voltages and field intensity were increased. The inactivation mechanism of the studied enzymes under PEF may be explained in terms of configuration changes due to the electrostatic nature of the enzymes as proteins. Enzymes are stabilized by weak non-covalent forces such as hydrogen bonds, electrostatic forces, van der Waals forces, hydrophobic interactions, and internal salt bridges (Price and Stevens, 1991). A change in the magnitude of any of these unions could cause denaturation. The application of high electric field pulses might have affected the forces involved in maintaining the three-dimensional structure (secondary, tertiary, and quaternary structure) or conformation of the globular protein.

PEF-treated enzyme solutions showed no significant changes in activity after 24 and 48 hr of storage at 4°C, suggesting a permanent inactivation of the enzymes when exposed to PEF.

It is important to mention that the temperature effect on the enzymes, or the degree of enzymes inactivation by heat treatment, is directly related to the water

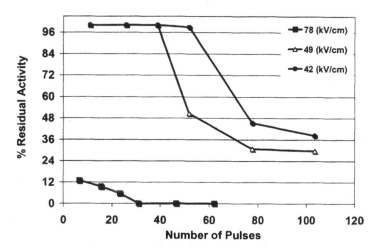

Figure 9.2 Effect of electric field and number of pulses on the residual enzymatic activity of Protamex™: Alcalase™ mixture after PEF treatment.

content of the system (Nagodawithada and Reed, 1993). Enzymes in a dry or semi-moist food system tend to be more heat stable. In this study, low protein percentage dispersions were submitted, so the water facilitated the unfolding of enzymes and faba protein during thermal denaturation. This unfolding explains, in part, the loss of enzyme activity and protein solubility, while the effect of pH on enzyme catalysis caused the ionization of substrate or enzyme. This can affect substrate binding or transformation to a product directly, or can affect enzyme conformational stability.

SOLUBILITY OF PROTEIN CONCENTRATES

Figure 9.3 shows the solubility profile (PSI) of the HPC after PEF treatments at three different electric fields and different numbers of pulses. The decrease in the PSI at high numbers of pulses can be correlated with detected changes in color and texture of the treated suspensions. In the case of the PC suspensions, their color and texture changed after the application of 104 pulses using 42 or 49 kV/cm, as well as with 23 pulses of 78 kV/cm. Such changes could be due to the degradation of the carbohydrates present on the concentrate and/or faba bean protein denaturation. Furthermore, HPC treated with a higher number of pulses at intermediate electric fields (181 pulses at 42 or 49 kV/cm vs. 78 pulses at 49 kV/cm) started to turn black (the original color of the suspension was a light brown) and began to clump. At high electric fields protein denaturation and darkening was present after 23 pulses.

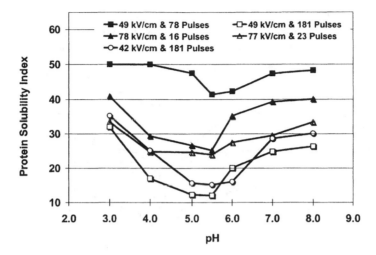

Figure 9.3 Effect of PEF treatment conditions on the protein solubility index of protein hydrolysates at different pH values.

Figure 9.4 shows the faba PC (without hydrolysis) solubility as a function of the suspension pH (range 3.0–8.0). Comparison of the findings indicates that the inactivation methods affected the PSI of the protein concentrates. Figure 9.5 shows the solubility profiles found for the PC (control without hydrolysis or treatment) and HPC treated with pH, temperature, and PEF (78

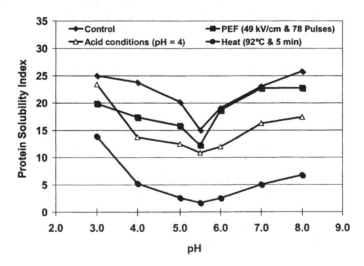

Figure 9.4 Effect of enzymatic inactivation method on the protein solubility index of protein concentrates at different pH values.

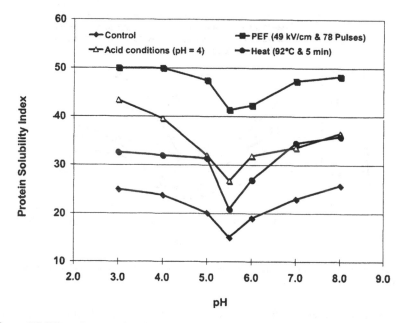

Figure 9.5 Effect of enzyme inactivation method on the protein solubility index of protein hydrolysates at different pH values.

pulses at 49 kV/cm). The lower solubility of the HPC treated with thermal inactivation (~34% at pH 7), compared with the PEF treated ones (~48% at pH 7), suggests that denaturation of the proteins caused insolubility. This denaturation involved dissociation and transition of the native state to an unfolded denatured state without alteration of the amino acid sequence (Adler-Nissen, 1986; Boyle et al., 1997). After the quaternary structure was destroyed, and the protein molecules broke up into a several sub-units, the conversion into an insoluble form caused by polymerization began. This brings variations in hydration properties (Sheards et al., 1986).

WATER ABSORPTION CAPACITY (WETTABILITY: WAC)

Figure 9.6 illustrates the behavior of PC (control) and HPC WAC, as a function of time (s). The time to reach equilibrium was different for all samples and was strongly dependent on the method and equipment used. The maximum amount of water absorbed (mL water/gram of protein) in the PC and HPC was similar to that reported by other research groups (e.g., Pilosof et al., 1985).

Table 9.2 shows the WAC and solubility at pH 5.5 of the PC and HPC products. Although it is not always possible to define a correlation between the

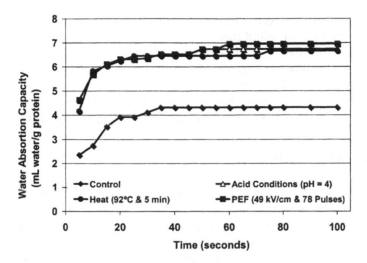

Figure 9.6 Effect of enzyme inactivation method on the water absorption capacity of the protein hydrolysates.

WAC and other functional properties, the results obtained by this study indicate that concentrates with low WAC also have a low solubility.

CONCLUSION

Pulsed electric field treatment has been shown to be an effective method for minimizing or completely inactivating commercial enzymes (Protamex[TM] and Alcalase[TM]). The Protamex[TM] and Alcalase[TM] mixture activity was reduced by 70% under the PEF application of 78 pulses of 49 kV/cm; in this case, the remaining activity did not increase over time. The inactivation mechanism of

TABLE 9.2. **Water Absorption Capacity (WAC) and Solubility (PSI) of Faba Bean Products.**

Product	WCA	PSI at pH 5.5
Protein concentrate (PC: control)	4.31	15.09
Hydrolyzed protein concentrate stabilized with heat treatment (92°C and 5 min)	6.66	20.95
Hydrolyzed protein concentrate stabilized with acidification treatment (pH drop to 4)	6.73	26.75
Hydrolyzed protein concentrate stabilized with PEF treatment (49 kV/cm and 78 pulses)	6.94	41.41

the studied enzymes under PEF may be explained in terms of configuration changes and loss of helical structure due to the electrostatic nature of the enzymes. Furthermore, PEF allowed the production of stabilized HPCs suitable for producing nutritional supplements, with better solubility characteristics than those stabilized with acidification and heat treatments. These results encourage broader studies on the applications of PEF as a nonthermal technology to control enzymatic activity, as well as in the area of food product development.

ACKNOWLEDGEMENTS

We express our sincere thanks to the Departamento de Química, Vicedecanatura de Bienestar Universitario—Facultad de Química, Vicerrectoría de Bienestar Universitario; Universidad Nacional de Colombia, CINDEC for their support to Lilian Palomeque in her master's studies; to the Universidad Nacional Autónoma de México (UNAM) and CONACyT (México) for supporting M. Marcela Góngora-Nieto's doctoral studies at WSU; and to COLDANZIMAS for supplying the enzymes.

REFERENCES

Adler-Nissen, J. 1976. Enzymatic hydrolysis of proteins for increased solubility. *J. Agric. Food Chem.* 24(6):1090–1093.

Adler-Nissen, J. 1986. *Enzymatic Hydrolysis of Food Proteins.* Elsevier, New York. pp. 55, 74, 76–78.

AOAC, 1990. Official methods of analysis of the Association of Official Analytical Chemists. Washington, DC.

Baquero, C. J. and Bermúdez, A. S. 1997. Modificación de las propiedades funcionales del concentrado proteico de haba mediante tratamiento enzimático. B.S. thesis, Departamento de Química, Universidad Nacional de Colombia, Bogotá, Colombia.

Barbosa-Cánovas, G. V., Gongora-Nieto, M. M., Pothakamury, U. R., and Swanson, B. G. 1999. *Preservation of Foods with Pulsed Electric Fields.* Academic Press, San Diego, CA.

Barbosa-Cánovas, G. V., Pothakamury, U., Palou, E., and Swanson, B. G. 1998. *Nonthermal Preservation of Foods.* Marcel Dekker, Inc., New York. pp. 53–65.

Boyle, J. I., Alli, I., and Ismail, A. A. 1997. Use of differential scanning calorimetry and infrared spectroscopy in the study of thermal and structural stability of α-lactoalbumin. *J. Agric. Food Chem.* 45:116–1125.

Camacho, F., González-Tello, P., and Guadix, E. M. 1998. Influence of enzymes, pH and temperature on the kinetics of whey protein hydrolysis. *Food Science and Technology International.* 4:79–84.

Cantoral, R., Fernandez-Quintela, A., Martinez, J. A., and Macarulla, M. T., et al. 1995. Estudio comparativo de la composición y el valor nutritivo de semillas y concentrados de proteínas de leguminosas. *Archivos Latinoamericanos de Nutrición.* 45(3):242–248.

De Kanterewicz, B. E., Pilosof, A. M. R., and Bartholomai, G. B. 1989. A simple method determining the spontaneous oil absorption capacity of proteins and the kinetics of oil uptake. *JAOCS.* 66(6):809–812.

Elizalde, B. E., Pilosof, A. M. R., and Bartholomai, G. B. 1996. Empirical model for water uptake and hydration rate of food powders by sorption and Baumann methods. *Journal of Food Science.* 61(2):407–409.

Frokjaer, S. 1994. Use of hydrolysates for protein suplementation. *Food Technology.* 48(10): 86–88.

Ho, S. V., Mittal, G. S., and Cross, J. D. 1995. Effects of high field electric pulses on the activity of selected enzymes. *Journal of Food Engeneering.* 9:210–214.

Mahmoud, M. I. 1994. Physicochemical and functional properties of protein hydrolysates in nutritional products. *Food Technology.* 48(10):89–95.

McCurdy, S. M. and Knipfel, J. E. 1990. Investigation of faba bean protein recovery and application to pilot scale processing. *Journal of Food Science.* 55(4):1093–1101.

Montilla, J. 1994. Importancia agronómica y nutricional de las leguminosas. *Archivos Latinoamericanos de Nutrición.* Supplement 1, 44(4):44–49.

Mordecay, D. L. and Bermúdez, A. S. 1994. Obtención de productos proteicos a partir de semillas de habas (*Vicia faba*). M. S. thesis, Universidad Nacional de Colombia, Departamento de Química, Bogotá, Colombia.

Morr, C. V., German, B., Kinsella, J. E., Regenstein, J. M., Van Buren, J. P., Kilara, A., Lewis, B. A., and Mangino, M. E. 1985. A collaborative study to develop a standardized food protein solubility procedure. *Journal of Food Science.* 50:1715–1718.

Mosquera, M., Martínez, M. J., and Bermúdez, A. S. 1996. Preparación de un hidrolizado proteico de haba (*Vicia faba*) que pueda ser incluido en un producto alimenticio. B.S. thesis, Departamento de Farmacia, Universidad Nacional de Colombia, Bogotá, Colombia.

Nagodawithada, T. and Reed, G. 1993. *Enzymes in Food Processing.* Academic Press, San Diego, CA. pp. 39–58.

Niño, M., Heredia, R. D., and Bermúdez, A. S. 1996. Preparación de un hidrolizado proteico de haba (*Vicia faba*) que pueda ser incluido en una bebida. B.S. thesis, Departamento de Farmacia, Universidad Nacional de Colombia, Bogotá, Colombia.

Palomeque, L. A. and Bermúdez, A. S. 1995. Efecto de la modificación enzimática y del método de deshidratación sobre la solubilidad del aislado proteico de haba (*Vicia faba*). B.S. thesis, Departamento de Química, Universidad Nacional de Colombia, Bogotá, Colombia.

Palomeque, L. A. and Bermúdez, A. S. 1997. Evaluation of factors involved in the protein solubility index of faba bean (*Vicia faba*) protein isolate. CoFE Poster Sessions at AIChE Annual Meeting, USA.

Pedersen, B. 1994. Removing bitterness from protein hydrolysates. *Food Technology.* 48(10): 96–99.

Pilosof, A. 1995. Desarrollo de concentrados de proteína de soya de alta funcionalidad. *Programa Iberoamericano de Ciencia y Tecnología para el Desarrollo, Subprograma XI-Tratamiento y Conservación de Alimentos*, RIARE.

Pilosof, A., Boquet, R., and Bartholomai, G. B. 1985. Kinetics of water uptake by food powders. *Journal of Food Science.* 50:278–282.

Price, N. C. and Stevens, I. 1991. Enzymes: Structure and functions. Vol. I–II. In: *Food Enzymology*, P. F. Fox, Ed. Elsevier, London. pp. 1–25.

Qin, B. L., Vega-Mercado, H., Pothakamury, U., Barbosa-Cánovas, G.V., and Swanson, B.G. 1995. Application of pulsed electric fields for inactivation of bacteria and enzymes. *Journal of the Frankin Institute.* 332A:209–220.

Schmidl, M. K., Taylor, S. L., and Nordlee, J. A. 1994. Use of hydrolysate-based products in special medical diets. *Food Technology.* 48(10):77–85.

Sheards, P. R., Fellows, A., Ledward, D. A., Mitchell, J. R. 1986. Macromolecular changes associated with the heat treatment of soya isolate. *Journal of Food Technology.* 21:55–60.

Vega-Mercado, H., Powers, J. R., Barbosa-Cánovas, G. V., and Swanson, B. G. 1995. Plasmin inactivation with pulse electric fields. *Journal of Food Science.* 60(5):1143–1146.

Zhang-Ying, L. and Yan, W. 1993. Effects of high voltage pulse discharges on microorganisms dispersed in liquid. *8th International Symposium on High Voltage Engineering*, Yokuhama, Japan. pp. 551–554.

CHAPTER 10

Inactivation of *Listeria innocua* and *Pseudomonas fluorescens* in Skim Milk Treated with Pulsed Electric Fields (PEF)

J. J. FERNÁNDEZ-MOLINA
E. BARKSTROM
P. TORSTENSSON
G. V. BARBOSA-CÁNOVAS
B. G. SWANSON

ABSTRACT

SURVIVAL curves of *L. innocua* and *P. fluorescens* exposed to pulsed electric fields (PEF) have exponential decaying form when plotted in linear co-ordinates, and follow linearity when plotted in semi-logarithmic coordinates. These survival curves were described by application of the model represented by Equation (2) of this chapter. This model worked best at electric fields ranging from 40–50 kV/cm, 40–100 pulses, and treatment times from 100–200 μs. The kinetic constants determined for *L. innocua* and *P. fluorescens* ranged from 2.467 μs^{-1} at 30 kV/cm to 0.131 μs^{-1} at 50 kV/cm for *L. innocua,* and from 4.470 μs^{-1} to 0.396 μs^{-1} for *P. fluorescens* at the same electric field intensities.

The viability of *Listeria innocua* and *Pseudomonas fluorescens* before and after treatment were assayed by counting colony-forming units (cfu). Logarithmic reduction of 2.7 and 2.6 cycles in *L. innocua* and *P. fluorescens* in raw skim milk were achieved when treated at electric field intensities of 30, 40–50 kV/cm with 30–100 pulses, 2 μs at a frequency of 4.0 Hz. A coaxial continuous treatment chamber, with electrode gaps of 0.63 cm, was selected for high intensity pulsed electric field (PEF) treatment. Effective lethality of PEF treatment was achieved at electric field intensities greater than 30 kV/cm and 40 pulses, treatment time 50–200 μs for both microorganisms. Raw skim milk was maintained at temperatures of 7, 25, or 28°C during PEF treatment.

INTRODUCTION

During the last 20 years, the inactivation of microorganisms and enzymes in food products by pulsed electric fields (PEF) has gained the interest of the scientific community and the food industry. This emerging technology is a unique method of food preservation in which the inactivation of microorganisms and enzymes is achieved in a very short time with little heating of the medium.

One of the earliest applications of electricity in food processing was the pasteurization of milk in the 1920s, with the "electro-pure process" (Vega-Mercado et al., 1999). Microbial inactivation occurred as a result of ohmic heating. The electro-pure process inactivates *Mycobacterium tuberculosis* and *Escherichia coli*. Getchell (1935, quoted by Palaniappan et al., 1990) reported that electric pasteurization not only destroyed some of the pathogenic bacteria commonly found in milk, but was also an effective safeguard against certain bacteria on which other methods of pasteurization had little or no effect.

Electric pasteurization of milk consisted of pumping the milk through a regenerative (heat exchange) coil, an electrical heating chamber, and a surface heat exchanger for cooling. The electrical chamber consisted of a vertical rectangular tube with opposable walls of carbon electrodes and heavy glass for insulation (Barbosa-Cánovas et al., 1999). A 220-V alternating current supply with a constant power of 15 kW was applied to the carbon electrodes, and raw milk preheated to 52°C was passed through the treatment chamber. The electric current passing through the milk in the treatment chamber raised the temperature to 71°C. The milk was then cooled to approximately 29°C and bottled or collected for cream separation (Barbosa-Cánovas et al., 1999). This process has not been used in the dairy industry since the 1960s (Getchell, 1935).

Castro et al. (1993) reviewed the literature on inactivation of microorganisms and enzymes with PEF and suggested that high intensity PEF is potentially the most important nonthermal pasteurization/sterilization food preservation technology available to replace or complement thermal processes. Microbial survival depends exponentially on applied field intensity, and the threshold-applied voltage is different among microbes. During the PEF process, lysis of microorganisms is caused by irreversible structural changes in biological membranes, leading to pore formation and destruction of the semipermeable barrier of the membrane. In a suspension of cells, an electric field causes potential differences across the membrane, inducing a sharp increase in membrane conductivity and permeability. Membrane destruction occurs when the induced membrane potential exceeds a critical value of 1 volt in many cellular systems, corresponding to an external electric field of approximately 10 kV/cm for *E. coli* (Castro et al., 1993).

Calderón-Miranda (1998) studied the inactivation of *Listeria innocua* suspended in skim milk treated with PEF, as well as the sensitization of PEF treated

L. innocua to nisin. Increasing reduction of the population of *L. innocua* in skim milk of 2-½ log cycles was observed at 30, 40, and 50 kV/cm. In addition 2, 2.7, and 3.4 log reduction cycles of *L. innocua* were achieved by exposure to 10 IU nisin/mL after 32 pulses with the same applied electric fields. A synergistic effect was observed in the inactivation of *L. innocua* as a result of exposure to nisin after PEF, compared to inactivation of *L. innocua* before exposure to nisin.

Reina et al. (1998) studied the inactivation of *Listeria monocytogenes* in pasteurized whole, 2%, and skim milk with PEF, and observed no significant differences in the inactivation of *L. monocytogenes* Scott A. A 1 to 3 log reduction of *L. monocytogenes* was achieved at a treatment temperature of 25°C. A 4 log reduction of *L. monocytogenes* was accomplished by increasing the treatment temperature to 50°C. The lethal effect of PEF was a function of the field intensity and treatment time. Ho et al. (1995) inactivated *Pseudomonas fluorescens* with PEF in various aqueous solutions of peptone, sucrose, xanthan gum, and sodium chloride under different operating conditions and fluid properties. The applied electric fields ranged from 10 to 45 kV/cm, with the number of pulses ranging from 10 to 100, with a pulse duration of 2 to 4 μs. It was observed that a pulsed electric field intensity of 10 kV/cm at 2 s intervals and 10 pulses was sufficient to achieve a microbial reduction greater than 6 log cycles.

MATHEMATICAL MODEL

Esplugas (1996) proposed a mathematical model to study the inactivation kinetics of microorganisms using PEF continuous and recirculation processes. A recirculation PEF operation is illustrated in Figure 10.1. First order reaction

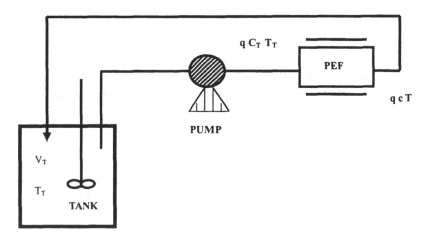

Figure 10.1 Schematic view of a recirculation PEF operation.

rate kinetics are hypothesized to inactivate microorganisms, assuming that the concentration of microorganisms (c_T) in the feed tank and at the entrance of the PEF chamber is the same (Figure 10.1). The reaction rate constant, r, with respect to the concentration of microorganisms, c (microorganisms/liter · second), is defined as:

$$r = -kc \tag{1}$$

where k (s^{-1}) is the reaction rate constant. A mass balance of microorganisms fed to the PEF chamber can be made, providing a stationary flow in the chamber where q (L/s) is the fluid flow rate and λ_r (L) is the volume of the PEF chamber, leading to the following expression:

$$q \ln \left(\frac{c}{c_T} \right) = -k\lambda_r t \tag{2}$$

Equation (2) is expressed as an exponential relationship between the concentration of microorganism/L, c, at the exit of PEF chamber and time, t(s):

$$c = c_T \exp(\theta k \, \lambda_{rt}/q) \tag{3}$$

A material balance around the tank can be made (Figure 10.1), assuming a perfect barrier and non-stationary conditions, leading to the following equation:

$$c_T = c_{T_0} \exp(-q/\delta_T (1 - \exp(-k\lambda_r/q)) \, t) \tag{4}$$

where λ_r/q and δ_T/q are the residence times in the PEF chamber and tank, respectively. δ_T (L) is the tank volume and c_{T_0} is the initial concentration of microorganisms.

Energy (E) from a high voltage d.c. power supply (Figure 10.2) stored in a capacitor and discharged through the food material, is given by Equation (5) (Barbosa-Cánovas et al., 1999):

$$E = 0.5^* C^* V^2 \tag{5}$$

where C is the capacitance (μF) and V is the charging voltage. Considering a pulse frequency of $f(L/s)$, the dissipated energy flow rate, Q (J/s), is denoted by:

$$Q = f^* 0.5 C^* V^2 \tag{6}$$

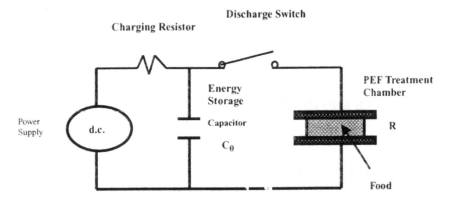

Figure 10.2 Typical pulser configuration for high voltage PEF.

The dissipated energy, Q (J/s), is discharged through the fluid flow, q (L/s), which passes through the PEF chamber, with density, ρ (kg/L), and specific heat, C_P (J/kg°C). An energy balance of the fluid food in the PEF chamber, assuming adiabatic conditions, leads to the following equation:

$$q * \rho * C_P (T - T_r) = rQ \qquad (7)$$

A material balance in the feed tank in a non-stationary environment can be made assuming adiabatic conditions and a model with a perfect barrier. Thus, the temperature variation, T_T (°C), in the tank with respect to time, t, is given by:

$$T_T = T_{T_0} + \frac{rQ}{\delta_T C_P \rho} t \qquad (8)$$

where T_{T_0} is the initial temperature in the tank (°C) with a 40% accumulation of the dissipated energy in the tank ($r = 0.4$).

The two strains of microorganisms were selected because they are part of the microflora of raw milk. Neither of the microorganisms are pathogenic, but the presence of *L. innocua* in milk may indicate the presence of pathogenic *L. monocytogenes* (Calderón-Miranda, 1998). On the other hand, *Pseudomonas* spp. has been observed in raw milk stored in silos, on the order of 70% of the total flora present (Marth and Steele, 1998). *Pseudomonas* spp. are well adapted to survival in the milk-processing environment regardless of the production location, and they are able to adhere strongly to the surface of milk processing equipment (Smithwell and Kallasapathy, 1995). *Pseudomonas* spp. are the major type of spoilage bacteria in pasteurized milk at the end of its shelf life when stored at the recommended temperature of 4°C.

The objective of this research was to study the inactivation of *Listeria innocua* and *Pseudomonas fluorescens* by PEF treatments, and to apply a simple mathematical model that accurately predicts the PEF inactivation of these microorganisms.

MATERIALS AND METHODS

SKIM MILK

Raw milk was provided by the Washington State University Creamery (Pullman, WA). Milk fat was separated using a Model 614 separator (DeLaval, Sweden) to reduce the fat content from 3.7% to approximately 0.2%. An inoculum of 2 mL of each microorganism was added to 3 L of sterile skim milk (7°C) to get initial concentrations of 1.2×10^9 and 3.8×10^7 cfu/mL of *L. innocua* and *P. fluorescens*, respectively, for PEF treatment. Before and after each PEF treatment, serial dilutions of *L. innocua* and *P. fluorescens* were performed in sterile 0.1% peptone (DIFCO Laboratories, Detroit, MI). Dilutions were plated on tryptic soy agar (BIOPRO) enriched with 0.6% yeast extract (TSAYE), and on *Pseudomonas* agar F (DIFCO Laboratories, Detroit, MI), for *L. innocua* and *P. fluorescens*, respectively.

MICROBIAL PREPARATION

Listeria innocua (ATCC 51742, Rockville, MD) and *Pseudomonas fluorescens* (ATCC 17926) were used in this study. Bacteria were grown according to ATCC procedure. Tryptic soy broth (DIFCO) enriched with 0.6% yeast extract (TSBYE) was used as the growth medium for *L. innocua*. One milliliter of frozen culture was thawed and inoculated in 50 mL of TSBYE with continuous agitation at 190 rpm in a temperature controlled shaker (Model BSB-332A-1, GS Blue Electric, Blue Island, IL) at 37°C for 18 hr to reach early stationary phase. The growth of bacteria cells was followed by absorbance at 540 nm in a UV-light spectrophotometer (Calderon-Miranda, 1998). Cells of *L. innocua* were harvested by centrifugation of the culture solution at 5500 rpm for 10 min at 10°C using a centrifuge (Beckman J-21C, NY). *L. innocua* cells were washed with the chilled TSBYE, centrifuged (5000 rpm, 10°C, 5 min), suspended in 60 mL TSBYE, and stored at −70°C with 1 mL of 20% glycerol until further use.

Cells of *P. fluorescens* (ATCC 17926) were grown in tryptone soya yeast extract (TSYE) at 25°C for 48 hr to provide a concentration of cells/mL of approximately 3.8×10^7. Stationary phase and concentration were measured by spectrophotometer at 540 nm, and a calibration growth curve was made to convert optical density to cell concentration. *Pseudomonas* cells were harvested from the broth using a centrifuge (Beckman model J-21C, NY) at 5380 g, 4°C for 10 min (Ho et al., 1995). Harvested cultures were stored at −70°C with 1 ml of 20% glycerol until further use.

PULSED ELECTRIC FIELD

A coaxial treatment chamber (Figure 10.3), with a 28.7 cm³ capacity and electrode gaps of 0.63 cm, was used for high intensity pulsed electric field (PEF) treatment. The intensity of the electric field and pulse waveform was determined with an oscilloscope (Hewlett Packard 54520A, Colorado Springs, CO), and the electric field was generated using a pilot plant size pulser manufactured by Physics International (San Leonardo, CA). For this study the frequency was 3.5 or 4 Hz and the flow rate was 500 mL/min. The flow rate of the skim milk was controlled with a rotary pump (Master Flex model 7518-00, Cole Parmer Instruments Co., Chicago, IL). A cooling coil immersed in an ethylene glycol

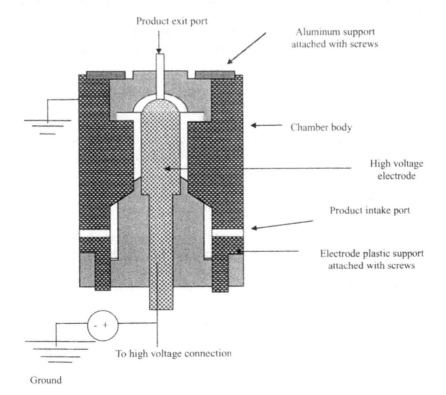

Figure 10.3 Configuration of PEF coaxial continuous treatment chamber used in this study.

bath was used to cool the skim milk at the entrance and outlet of the PEF chamber. The number of pulses given to the skim milk was determined from the following expression (Barbosa-Cánovas et al., 1999):

$$n = (v/f)^* F \qquad (9)$$

where n represents the number of pulses, v the volume of the treatment chamber (mL), f the flow rate of the milk (mL/s) and F the pulsing frequency (Hz).

An energy storage capacitor of 0.5 μF was used, and the input voltage was set at 30, 35, or 40 kV (Figure 10.2). The approximate maximum electric field intensity calculated from data gathered with the oscilloscope was 30, 40, or 50 kV/cm. A stepwise process was used to apply a set of 10 to 100 pulses to the skim milk for the inactivation of *L. innocua,* and 10 to 50 pulses were applied to the skim milk for the inactivation of *P. fluorescens.* The waveform used was exponential decay with a pulse length of ~2 μs. The PEF treatment temperatures ranged from 15°C to 28°C. The temperature at the entrance and exit of the treatment chamber was recorded with a digital thermometer (John Fluke, Everett, WA).

The maximum electric field strength, E (kV/cm), between the two coaxial electrodes of the PEF chamber was determined by the following expression

$$E = \frac{V}{r\left[\ln\left(\frac{R_2}{R_1}\right)\right]} \qquad (10)$$

where V (kV) is the input voltage, r is the radius where the electric field is measured ($r = R_1 - (R_2 - R_1)/2$), and R_1 and R_2 are the inner and outer radii in centimeters (Barbosa-Cánovas et al., 1999).

EXPERIMENTAL DESIGN

A two level factorial design was used to evaluate the effect of electric field strength (30, 40, or 50 kV/cm), number of pulses (30, 40, 50, and 100), on the inactivation of *L. innocua* and *P. fluorescens.* Two replicates were made for each variable combination and assays were replicated twice.

STATISTICAL ANALYSIS

Data was analyzed using Microsoft Excel 97 for Windows (1998), applying the least square method by minimizing the sum of squares. A regression analysis was performed on each set of data points to establish a relationship between experimental and predicted values of survivor fractions in linear and semi-logarithmic plots. The regression parameters were judged by the magnitude

TABLE 10.1. Survival Constants of *L. innocua* and *P. fluorescens* Exposed to Pulsed Electric Fields Estimated from the Fit of Equation (2) as a Model.

Organism	Number of Pulses	Electric Field (KV/cm)	k (1/μs)	r^2	Experimental r^2	Model r^2
L. innocua	0–50	30	2.467	0.260	0.176	1
	0–50	40	1.726	0.961	0.961	1
	0–100	50	0.131	0.973	0.945	1
P. fluorescens	0–30	30	4.470	0.525	0.296	1
	0–40	40	2.315	0.984	0.919	1
	0–50	50	0.396	0.998	0.808	1

of r^2 between -1 and 1. A significant level of $p = 0.05$ was established for significant difference between treatments. Comparison between treatments was made using a two sample t-test with equal variance at $p = 0.05$.

RESULTS AND DISCUSSION

THE MODEL

Schematic representation of survival curves of *L. innocua* and *P. fluorescens* exposed to pulsed electric fields are illustrated in Figures 10.4 to 10.9. Survival curves in Figures 10.4(a) to 10.9(a) represent the plots of the survival fraction against the treatment time generated with Equation (2). Survival curves in Figures 10.4(b) to 10.9(b) are the same plots in semi-logarithmic coordinates representing linearity of the model. The kinetics constants at different electric field intensities and number of pulses are depicted in Table 10.1. According to this model, the surviving organisms of *L. innocua* and *P. fluorescens* fall exponentially when plotted in linear coordinates, and approach linearity by plotting the ln (survival fraction) vs. treatment time.

PEF INACTIVATION OF *L. INNOCUA* AND *P. FLUORESCENS* IN SKIM MILK

Application of Equation (2) to the individual survival curves of *L. innocua* and *P. fluorescens* is given in Figures 10.4 to 10.9. The inactivation constants are listed in Table 10.1. A high and strong significant effect ($p = 0.05$) of the electric intensity on the inactivation of both microorganisms was observed at electric fields of 40–50 kV/cm. The regression coefficient of *L. innocua* ranged from 0.260–0.973, and ranged from 0.525–0.998 for *P. fluorescens*. No significant effect on the inactivation of *L. innocua* and *P. fluorescens* was observed at 30 kV/cm as shown in Table 10.1. This poor fit could be attributed

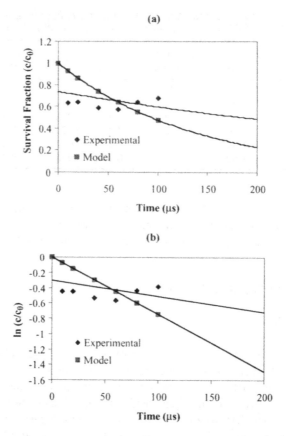

Figure 10.4 Inactivation of *L. innocua*, suspended in skim milk, by PEF treatments (30 kV/cm, 4 Hz, 2 μs): (a) in linear coordinates generated with Equation (2), and (b) same curve in conventional semi-logarithmic coordinates.

to the insufficient electric field needed to kill the organisms during the PEF treatment. Raso et al. (2000) demonstrated that higher lethality is obtained at high electric field. The lethality of the electric field progressively increased as the number of pulses increased. This phenomenon is manifested by a sharp decrease in the kinetic constant (k), for *L. innocua* it dropped from 2.467 μs^{-1} at 30 kV/cm to 0.131 μs^{-1} at 50 kV/cm. The same trend is observed for *P. fluorescens*, whose kinetic constant dropped from 4.470 μs^{-1} to 0.339 μs^{-1} with the same applied electric field intensity. The relationship between experimental survival fractions and predicted values is presented in Table 10.1. As judged by statistical criteria, the fit was highly satisfactory at high electric field intensities, with r^2 ranging from 0.176–0.945 for *L. innocua*, and from 0.296–0.919 for *P. fluorescens* (as illustrated in Table 10.1).

(a)

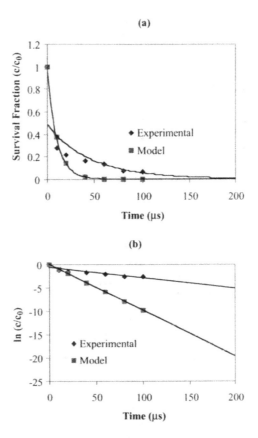

(b)

Figure 10.5 Inactivation of *L. innocua,* suspended in skim milk, by PEF treatments (40 kV/cm, 4 Hz, 2 μs): (a) in linear coordinates generated with Equation (2), and (b) same curve in conventional semi-logarithmic coordinates.

Survival curves of *L. innocua* and *P. fluorescens* at different electric field intensities, with a treatment frequency of 4 Hz and pulse width of 2 μs, are presented in Figures. 10.10 and 10.11. As observed in these figures, the inactivation kinetics of *L. innocua* and *P. fluorescens* are linear when the log 10 of the survival fraction was plotted against the number of pulses. The same trend was observed in Figures 10.4 to 10.7. The inactivation of both microorganisms progressively increased with increasing electric field and number of pulses. Figures 10.7 and 10.8 illustrate the inactivation of *L. innocua* (initial concentration 1.2×10^9) and *P. fluorescens* (initial concentration 3.8×10^7 cfu/mL) as a function of electric field intensities and number of pulses. A logarithmic reduction of 2.7 was attained for *L. innocua* with a peak electric field intensity of 50 kV/cm and 100 pulses, pulse duration of 2 μs, and 4 Hz frequency.

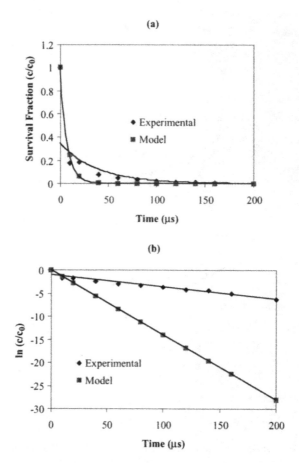

Figure 10.6 Inactivation of *L. innocua*, suspended in skim milk, by PEF treatments (50 kV/cm, 4 Hz, 2 μs): (a) in linear coordinates generated with Equation (2), and (b) same curve in conventional semi-logarithmic coordinates.

No significant differences ($p < 0.05$) were observed in the inactivation of *L. innocua* treated with electric field intensities of 30 or 40 kV/cm. The extent of inactivation of *L. innocua* increased when the applied electric field intensity was increased. Thus, permeabilization of the cell membrane of *L. innocua* occurred, altering the osmotic equilibrium and leading to swelling and eventual rupture of the cell membranes (Vega-Mercado et al., 1996). Calderón-Miranda (1998) achieved 2.4 log reduction of *L. innocua* treated with an electric field intensity of 50 kV/cm, 32 pulses, treatment time of 2 μs, and 3.5 Hz frequency. *P. fluorescens* was inactivated up to nearly 2.6 log cycles with an electric field intensity of 50 kV/cm, 50 pulses, treatment time of 2 μs, and 4 Hz frequency. A significant difference in the inactivation of *P. fluorescens* ($p = 0.05$) was not

(a)

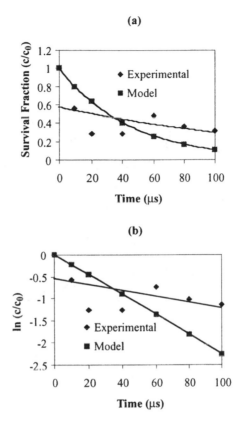

(b)

Figure 10.7 Inactivation of *P. fluorescens,* suspended in skim milk, by PEF treatments (30 kV/cm, 4 Hz, 2 μs): (a) in linear coordinates generated with Equation (2), and (b) same curve in conventional semi-logarithmic coordinates.

observed with electric field intensities of 30 or 40 kV/cm. With an increase in the electric field intensity, more energy is supplied to the cell suspension, therefore, more inactivation is attained. Ho et al. (1995) reported 6.4 log reduction of *P. fluorescens* when it was suspended in NaCl solution treated with PEF at 25 kV/cm, treatment time of 20 μs, and pulse duration of 2 μs. The inactivation was not dependent on the concentration of NaCl. Grahl and Märkl (1996) obtained 4.2 log reduction of *P. fluorescens* in UHT milk (1.5% to 3.5% fat) treated with 22 kV/cm, treatment time of 300 μs, and pulse duration of 15 μs. Although the PEF treatments used in the two experiments described above were different than the treatments applied in this research, the results confirm that inactivation of microorganisms with PEF is dependent on the electric field intensity, number of pulses, treatment time, pulse shape, frequency, temperature, electrical conductivity, homogeneity, and microflora.

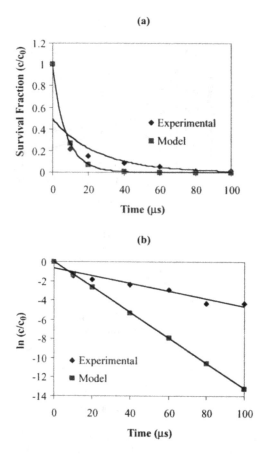

Figure 10.8 Inactivation of *P. fluorescens,* suspended in skim milk, by PEF treatments (40 kV/cm, 4 Hz, 2 μs): (a) in linear coordinates generated with Equation (2), and (b) same curve in conventional semi-logarithmic coordinates.

INPUT VOLTAGE AND HEATING OF SKIM MILK TREATED WITH PEF

The influence of input voltage (V) and thermal treatment on microbial inactivation with PEF is illustrated in Figure 10.12. The input voltage varied from 5000 V to 40,000 V with frequencies ranging from 0.5 Hz to 4 Hz. The temperature of the fluid ranged from 0°C to approximately 48°C within a selected range of input voltage and frequencies. The heat generated during PEF treatment is beneficial for the inactivation of microorganisms such as *L. innocua* and *P. fluorescens,* but abrupt increases in temperature during PEF treatment may not produce improved conditions. Therefore, refrigeration of the fluid at

(a)

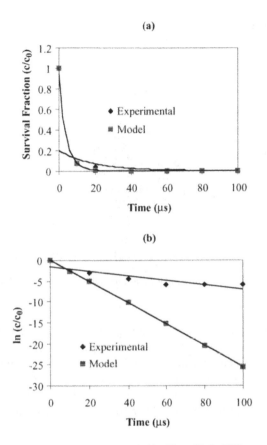

(b)

Figure 10.9 Inactivation of *P. fluorescens*, suspended in skim milk, by PEF treatments (50 kV/cm, 4 Hz, 2 μs): (a) in linear coordinates generated with Equation (2), and (b) same curve in conventional semi-logarithmic coordinates.

the entrance and exit of the PEF treatment chamber is required in order to avoid formation of air bubbles. Air bubbles induce arc discharge inside the treatment chamber, leading to damage of the electrodes. The temperature of the skim milk treated with PEF under the conditions described above did not exceed 28°C.

CONCLUSIONS

In conclusion, the model represented by Equation (2) accurately described the inactivation of *L. innocua* and *P. fluorescens* at different electric field intensities in the range of 40–50 kV/cm, number of pulses (40–100), and treatment time (50–200 μs). From a practical point of view, this mathematical model could be a good approach to use for analyzing PEF treatments to determine

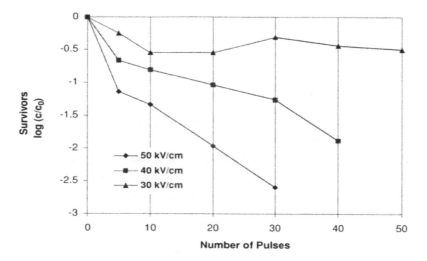

Figure 10.10 Survivor curves of *L. innocua* in skim milk, treated with PEF at different electric field intensities.

the rate of inactivation kinetics. Logarithmic reduction of *L. innocua* (2.7) and *P. fluorescens* (2.6) suspended in skim milk were achieved by a stepwise pulsed electric field treatment, adequate to inactivate the spoilage microorganisms in raw skim milk. The inactivation of *L. innocua* and *P. fluorescens* with PEF followed first order reaction kinetics. Significant differences were not observed when skim milk was treated with 30 or 40 kV/cm, 30 to 40 pulses, pulse width

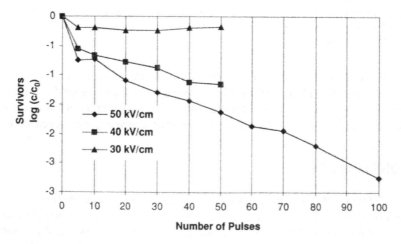

Figure 10.11 Survivor curves for *P. fluorescens* in skim milk, treated with PEF at different electric fields intensities.

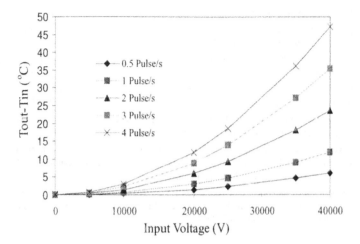

Figure 10.12 Influence of the input voltage, frequencies, and treatment temperatures of raw skim milk treated with PEF.

of 2 μs, and frequencies of 4 Hz. Increasing the input voltage during PEF treatment to 40 kV and a frequency of 4 Hz increased the treatment temperature to nearly 48°C.

In general, the inactivation of *L. innocua* and *P. fluorescens* during PEF treatment depends on electric field intensity, number of pulses, pulse width, frequency, pulse shape, temperature, electrical conductivity, homogeneity, and microflora.

REFERENCES

Barbosa-Cánovas, G. V., Góngora-Nieto, M. M., Pothakamury, U. R., and Swanson, B. G. 1999. *Preservation of Foods with Pulsed Electric Fields.* Academic Press, San Diego, CA.

Calderón-Miranda, M. L. 1998. Inactivation of *Listeria innocua* by pulsed electric fields and nisin. M.S. thesis. Washington State University, Pullman, WA.

Castro, A., Barbosa-Cánovas, G. V., and Swanson, B. G. 1993. Microbial inactivation in foods by pulsed electric fields. *J. Food Proc. Pres.* 17:47–73.

Esplugas, S. 1996. Pasteuritzacio continua i en recirculacio de fluids alimentaris per polsos electrics "PEF." Technical Report. Dept. of Biological Systems Engineering, Washington State University, Pullman, WA.

Getchell, B. E. 1935. Electric pasteurization of milk. *Agric. Eng.* 16(10):408–410.

Grahl, T. and Märkl, H. 1996. Killing of microorganisms by pulsed electric fields. *Appl. Microbiol. Biotechnol.* 45:148–157.

Ho, S. Y., Mittal, G. S., Cross, J. D., and Griffith, M. W. 1995. Inactivation of *Pseudomonas fluorescens* by high voltage electric pulses. *J. Food Science.* 60(6):1337–1343.

Marth, E. L. and Steele, J. L. 1998. *Applied Dairy Microbiology*. Marcel Dekker, Inc., New York.

Palaniappan, S., Sastry, S. K., and Richter. E. R. 1990. Effects of electricity on microorganisms: A review. *J. Food Proc. Pres.* 14:394–414.

Raso, J., Alvarez, I., Condón, S., Sala-Trepat, F. J. 2000. Predicting inactivation of *Salmonella senftenberg* by pulsed electric fields. *Innovative Food Sci. & Technol.* 1(1):21–29.

Reina, L. D., Jin, Z. T., Zhang, HQ. H., and Yousef. A. E. 1998. Inactivation of *Listeria monocytogenes* in milk by pulsed electric field: A research note. *J. Food Prot.* 61(9):120–106.

Smithwell, N. and Kallasapathy, K. 1995. Psychrotrophic bacteria in pasteurized milk: Problems with shelf life. *Australian Journal of Dairy Technol.* 50:28–31.

Vega-Mercado, H., Góngora-Nieto, M. M., Barbosa-Cánovas, G. V., and Swanson, B. G. 1999. Nonthermal preservation of liquid foods using pulsed electric fields. In: Rahman, M. S. *Handbook of Food Preservation*. Chapter 17. Marcel Dekker, Inc., New York.

Vega-Mercado, H., Pothakamury, U. R., Chang, F. J., Barbosa-Cánovas, G. V., and Swanson, B. G. 1996. Inactivation of *Escherichia coli* by combining pH, ionic strength, and pulsed electric fields hurdles. *Food Res. Inter.* 29(2):117–121.

CHAPTER 11

Inactivation of *Bacillus subtilis* Spores Using High Voltage Pulsed Electric Fields

Z. T. JIN
Y. SU
L. TUHELA
Q. H. ZHANG
S. K. SASTRY
A. E. YOUSEF

ABSTRACT

A suspension of *Bacillus subtilis* spores was treated by pulsed electric fields (PEF) in a bench scale, continuous system. Structural changes in PEF-treated spores were revealed by scanning electron microscopy (SEM). The release of dipicolinic acid (DPA) from spores was monitored during the PEF treatment. More than 95% of the *B. subtilis* spores were inactivated at an electric field strength of 30 to 40 kV/cm for 2 to 3 milliseconds. PEF treatment was most lethal to *B. subtilis* spores at 36°C. SEM micrographs showed that PEF-treated spores had structural changes similar to those of thermally inactivated spores. The DPA release was correlated with spore inactivation by PEF treatment.

INTRODUCTION

High voltage pulsed electric field (PEF) is a new technology that can be used to inactivate microorganisms without significant temperature increase. PEF has many advantages over conventional thermal preservation methods. Because food potentially can be pasteurized or sterilized by PEF at an ambient temperature, the color, texture, flavor, and nutrients of food are better preserved. Therefore, pulsed electric field pasteurization or sterilization is a potentially useful food processing technology (Zhang et al., 1995a).

Many researchers have demonstrated that PEF has lethal effects on microorganisms (Sale and Hamilton, 1967, 1968; Gupta and Murray, 1988; Mizuno and Hori, 1988; Jayaram et al., 1992; Zhang et al., 1994a, 1994b, 1995b).

Most of the earlier studies were carried out using only vegetative cells as target microorganisms. Because of their rigid structures, bacterial spores can survive harsh environments for a long period of time (Setlow, 1995). Limited studies have been done to inactivate bacterial spores by PEF. Some of these studies showed that bacterial spores are extremely resistant to PEF treatment (Hamilton and Sale, 1967; Yonemoto et al., 1993). Hamilton and Sale (1967) applied electric fields up to 30 kV/cm to spores of *Bacillus cereus* and *Bacillus polymyxa,* but the treatment did not affect viability of the spores. However, the authors reported that the electric fields had inactivation effects when the spores germinated. Yonemoto et al. (1993) treated spores of *Saccharomyces cerevisiae* and *Bacillus subtilis* with a pulsed electric field of 5.4 kV/cm. Their results indicated that 90% of yeast spores were inactivated by PEF, and that there was no viability change in bacterial spores. The scanning electron micrographs showed the PEF-treated yeast spores with many small holes on their surfaces, while structures of *B. subtilis* spores before and after PEF treatment were indistinguishable (Yonemoto et al., 1993). However, scanning electron micrographs revealed that there were some cracks on the surface of bacterial spores, and that the spores lost their viability after PEF treatment.

Dipicolinic acid (DPA), or pyridine 2,6-dicarboxylic acid, is a compound unique to bacterial spores and is involved in the thermal resistance of spores. When the integrity of the spore coats is compromised, DPA is released from the spore core into the surrounding environment. Release of DPA has been previously noted in *B. stearothermophilus* spores treated with ultrasonication (Palacios et al., 1991). PEF treatment damages spore coats and it is believed that DPA would be released from treated spores.

The purpose of this study was to evaluate the effectiveness of PEF treatment on the inactivation of *Bacillus subtilis* spores. The inactivation of spores was monitored by cell viability measurements, inspection of SEM micrographs, and assay for DPA release.

MATERIALS AND METHODS

PREPARATION OF *BACILLUS SUBTILIS* SPORES

Bacillus subtilis (OSU 872) spores were chosen as the target microorganisms. The vegetative cells of *B. subtilis,* at late logarithmic growth phase (9 to10 hr culture in tryptose broth, $O.D_{600} = 1.0$), were inoculated into bottle slants (containing nutrient agar $+ 0.03\%$ $MgSO_4$) and incubated at 37°C for 4 days. The spores were collected from the slant surface with chilled, sterile distilled water. For a mass spore production, the collected spore suspension was heated at 80°C for 20 min. A portion of heated spores (1 mL) was transferred to a large sporulation medium slant in a 150 mL dilution bottle, and incubated at 37°C for 4 to 6 days. Spores were washed off the agar surface with cold (ca. 0°C), sterile distilled water and combined in one flask. The spore suspension was centrifuged at 17,000 × g for 10 min at 4°C. There were two phases in the pellet. By microscopic examination, it was found that the upper phase contained spores with exosporium, and the lower phase contained matured spores. Therefore, the upper phase of the pellet was discarded. The remaining pellet was washed by resuspending into a saline solution followed by centrifugation at 17,000 × g for 10 min at 4°C. The washing procedure was repeated until a pure spore pellet was obtained. The final spore suspension was reconstituted with water to 10^8 cfu/mL and stored at 4°C before use.

PEF TREATMENT SYSTEM

A 15 kV high voltage power supply (Model 1450-4, Cober Electronics, Inc., Stamford, CT) was connected to a high voltage pulse generator (Model 2829, Cober Electronics, Inc., Stamford, CT) that can generate various waveform pulses. The pulse duration time and frequency were selected by a pulse trigger generator, and indicated by a two channel digital oscilloscope (Model TDS 320, Tektronix, Beaverton, OR). Since the square waveform is more effective than the exponential decay for inactivating microorganisms, as shown by Zhang et al. (1994b), the square waveform was selected for all the tests conducted in this study.

A continuous flow treatment chamber was designed to convert the high voltage pulses into high intensity pulsed electric fields. Inside the chamber, a converged electric field (valid treatment zone) formed between two liquid electrodes adjacent to the stainless steel electrodes. The two stainless steel electrodes were connected to the high voltage pulse generator and the ground, respectively. Two chambers, consisting of two 2 mm × 3 mm ($d \times l$) treatment zones, were connected in parallel in terms of the electric power connections. A schematic diagram of the PEF treatment system is illustrated in Figure 11.1.

Except where indicated, the treatment medium was an aqueous solution containing 3.42 mM NaCl and 1.14 mM L-alanine. The medium had conductivity of 1×10^{-4} S/cm and pH of 6.8 at 22°C. A gear pump (Micropump Inc., Vancouver, WA) carried the treatment fluid through the system and controlled the flow rate of the fluid. A water bath (Neslab, Newington, NH) was used to control the system temperatures. A stainless steel coil was placed before the PEF treatment chamber and immersed in the waterbath to control the PEF treatment temperature. A digital thermometer (Fluke 52 K/J, Everett, WA) with two thermocouples was used to measure the sample temperatures at the inlet and outlet of the system. The outlet temperature of the fluid was defined as

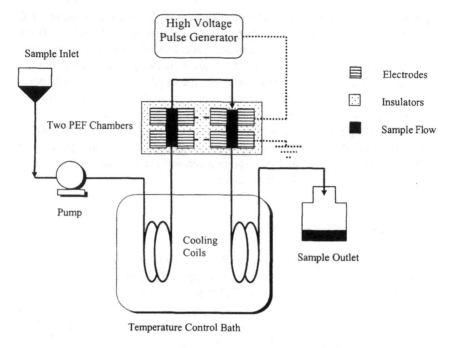

Figure 11.1 Pulsed electric fields treatment setup.

the treatment temperature. Treatment time (t) was calculated by Equation (1),

$$t = n \bullet \tau_c \bullet f \bullet d/v \tag{1}$$

where n is the number of treatment chambers, τ_c is effective pulse width (s), f is repetition frequency (Hz), d is distance between two electrodes (cm), and v is the average velocity of flow inside treatment chamber (cm/s).

TREATMENT OF *B. SUBTILIS* SPORES WITH PEF

A suspension of *B. subtilis* spores was prepared containing 3.42 mM NaCl and 1.14 mM L-alanine. The suspension was circulated immediately into the PEF treatment system. A control sample was taken at the inlet without PEF treatment. Samples also were taken at different treatment times.

ENUMERATION OF *B. SUBTILIS* SPORES

Counts of *B. subtilis* with and without heat activation were determined. Spore suspension (5 mL) was heated at 80°C for 20 min. Heated or unheated suspension was serially diluted in 0.1% peptone water, and surface plated in duplicate onto tryptose agar. Plates containing 20 to 200 colonies were counted.

EXAMINATION OF SPORES BY SCANNING ELECTRON MICROSCOPY

A PEF-treated spore suspension (35 mL) was centrifuged, and the pellet was spread onto a metal stub and dried in a laminar-flow hood for two hours. Untreated spores and thermally inactivated spores (121°C for 20 min) were also prepared for comparison. Samples were sputter coated and examined with a scanning electron microscope (JEOL JSM-820, Tokyo, Japan) at 20,000 × magnification.

MEASUREMENT OF DIPICOLINIC ACID RELEASE

The amount of DPA released from spores during PEF treatment was assayed by the method of Janssen et al. (1958), as modified by Rotman and Fields (1968). PEF-treated spore suspension was centrifuged and DPA concentration was determined in the supernatants. To determine the amount of unreleased DPA, the spore pellet was resuspended in a cold saline solution (0.02% NaCl), autoclaved for 15 min at 121°C, and separated by centrifugation. The supernatant was then analyzed for the concentration of DPA released from the pellets by heat.

The colorimetric assay for quantitative measurement of DPA was rapid and yielded a reproducible standard curve. Often the concentration of DPA in the supernatant of PEF-treated spores was not easily detected using colorimetric assay. In these situations, the difference in DPA concentration in spore pellets from PEF-treated and untreated samples was used. These data were used to estimate the concentration of DPA released from the spores, due to PEF treatment, by comparing the DPA concentrations from the treated samples to that present initially. The data were normalized for graphic presentation.

RESULTS AND DISCUSSION

PULSE DURATION AND FREQUENCY

Increasing pulse duration time from 3 to 12 μs, and concomitantly decreasing frequency from 2000 to 500 Hz, resulted in survival rates of *B. subtilis* spores that were not significantly different ($P > 0.1$) (Figure 11.2). Inactivation of spores depended on total treatment time irrespective of pulse duration-frequency combination used.

TREATMENT TEMPERATURE AND MEDIUM

The inactivation of spores by PEF was tested at 20, 30, 36, 40, and 50°C. Maximum inactivation occurred at 30 to 36°C (Figure 11.3). Sample temperature

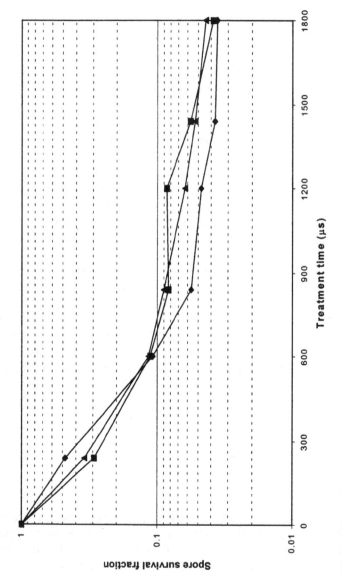

Figure 11.2 Inactivation of *Bacillus subtilis* spores at 36°C, electric field of 30 kV/cm, and different pulse durations and frequencies: (♦) 3 μs, 2000 Hz; (■) 6 μs, 1000 Hz; (▲) 12 μs, 500 Hz.

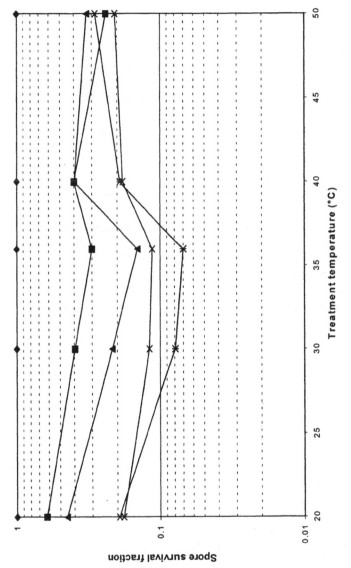

Figure 11.3 Inactivation of *Bacillus subtilis* spores at a frequency of 1500 Hz, electric field of 30 kV/cm, pulse duration of 2 μs, and different temperatures and treatment times. Treatment time (μs): (♦) none; (■) 270; (▲) 540; (×) 1080; (∗) 1620.

during PEF treatment influences inactivation rates differently when vegetative cells, insteady of bacterial spores, are used. Previous studies have shown that inactivation of vegetative cells increases with the increase in PEF treatment temperature (Dunn and Pearlman, 1987; Jayaram et al., 1993; Zhang et al., 1994a).

A germination control experiment (without PEF treatment) was performed at 36°C using a spore suspension similar to that used in PEF treatment experiments. Count of spores did not change appreciably during 48 min of incubation at 36°C; this indicates that spores did not germinate during the PEF treatment at that temperature.

Previous studies, however, indicate that germination of *Bacillus* spores was maximum at a temperature range from 30°C to 36°C (Levinson and Hyatt, 1970), and was inhibited at 49°C (Wax and Freese, 1968). Therefore, spores in the current study were at the most vulnerable state when they were exposed to the electric fields, since the treatment was done at temperatures most optimum for spore germination.

B. subtilis spores were suspended in an aqueous solution containing 3.42 mM NaCl with or without 1.14 mM L-alanine. Rate of inactivation was greater with alanine-containing medium than that containing NaCl only ($P < 0.01$) (Figure 11.4). L-alanine enhances germination of many strains of *B. subtilis* (Wax et al., 1967), but the amino acid did not cause the germination of the spores used in this experiment, as indicated earlier. However, combination of L-alanine and optimum temperature for spore germination may have caused physiological changes in spores, and increased their susceptibility to the PEF treatment.

ELECTRIC FIELD INTENSITY AND TOTAL PEF TREATMENT TIME

Spores were treated with PEF at three electric fields (30, 37, and 40 kV/cm), while the other parameters (pulse frequency, pulse duration, and treatment temperature) were kept unchanged. Results (Figure 11.5) are consistent with what has been reported on vegetative cells by other researchers (Sale and Hamilton, 1967; Hulsheger and Niemann, 1980; Hulsheger et al., 1981; Matsumoto et al., 1991; Jayaram et al., 1992). As the electric field intensity increased, the inactivation rate of spores increased. The maximum PEF treatment (40 kV/cm for 3500 μs), caused 98% spore inactivation. Statistical analysis to compare different data points (Figure 11.5) suggests that the total PEF treatment time has a significant effect on the inactivation of bacterial spores ($P < 0.01$).

SEM EVALUATION

Comparing the SEM micrographs (Figure 11.6) of PEF-treated and untreated *B. subtilis* spores revealed distinguishable structural differences between the two types of spores. Untreated spores have entirely smooth surfaces. After

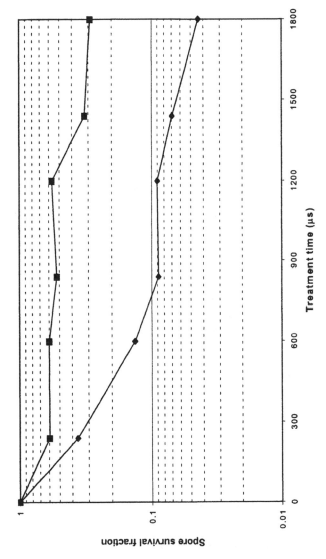

Figure 11.4 Inactivation of *Bacillus subtilis* spores in two treatment media at 36°C, electric field of 30 kV/cm, frequency of 1000 Hz, and pulse duration of 6 µs: (♦) 0.02% NaCl + 0.01% L-alanine (ALA); (■) 0.02% NaCl solution.

175

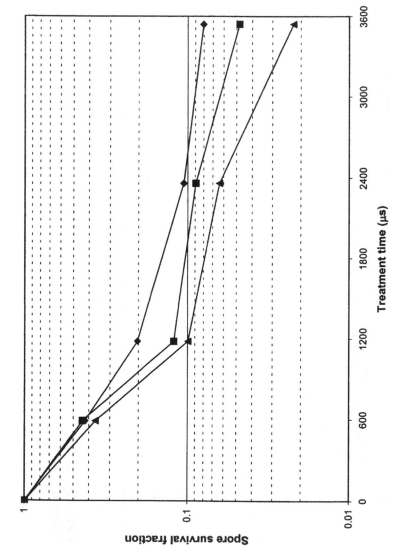

Figure 11.5 Inactivation of *Bacillus subtilis* spores at 36°C, frequency of 2000 Hz, pulse duration of 6 μs, and different electric field strengths: (♦) 30 kV/cm; (■) 37 kV/cm; (▲) 40 kV/cm.

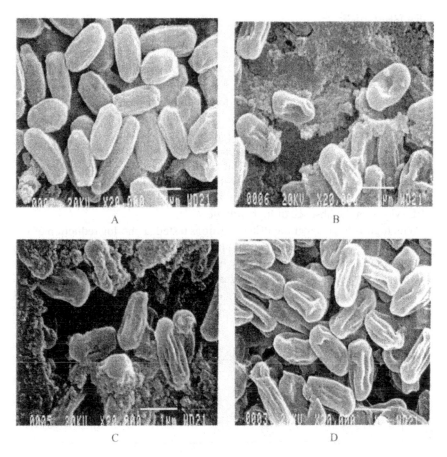

Figure 11.6 Scanning electron micrograph of *Bacillus subtilis* spores: (a) untreated; (b) treated with 100 pulses, 6 μs/pulse, and 30 kV/cm electric field; (c) treated with 600 pulses, 6 μs/pulse, and 30 kV/cm electric field; (d) heated at 121°C for 20 minutes.

PEF treatment, the spores shrank and many wrinkles formed on their surfaces. The shrinkage and wrinkles may indicate that the spores were inactivated. The SEM micrograph of heat inactivated spores [Figure 6(d)] illustrates similar structural changes. Compared to spores that were treated with 100 pulses, spores treated with 600 pulses had more and deeper wrinkles, and greater similarity to thermally inactivated spores. Plate count data showed that the 100 pulse treated sample had more surviving spores than the 600 pulse treated sample. The differences of the structure and survivability of 100 and 600 pulse treated spores might be explained by the reversible and irreversible electric breakdown theory (Zimmermann, 1986). The less pulse treated spores are better able to recover and remain viable (reversible), while the more pulse treated spores are deeply damaged and lose their viability (irreversible).

DPA RELEASE

PEF treatment caused a reduction in the number of viable spores, and the concomitant release of DPA (Figure 11.7). When the spore suspension was run through the PEF system in the absence of high voltage, the number of viable spores remained virtually unchanged after a time equivalent to 2000 μs of PEF treatment (Figure 11.7). By measuring the concentration of DPA remaining in the spore pellets of these samples, a small decline in DPA concentration in the spore pellets was observed between time 0 to 500 μs (equivalent run time), and the concentration remained unchanged from 500 to 1000 μs (equivalent run time) (Figure 11.7). These data suggest that there is little effect on the spore viability and DPA concentration when the spore suspension runs through the PEF system in the absence of high voltage.

With high voltage under the PEF conditions tested, a one-log reduction of viable spores was typically seen (Figure 11.7). During PEF treatment, the concentration of DPA in the spore pellet increased slightly after 500 μs, but generally decreased after 1000 and 2000 μs (Figure 11.7). After 500 μs of treatment, the spore coats may have sustained some damage, such as pores caused by the PEF treatment. While these pores may not penetrate all of the spore coats through

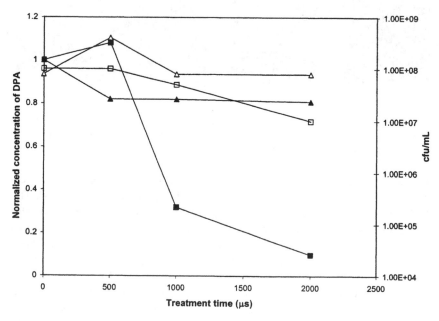

Figure 11.7 The effect of PEF treatment on viable spore counts and DPA concentration. Spore count: (△) untreated, (□) PEF treated. Normalized concentration of DPA: (▲) untreated, (■) PEF treated.

to the spore core, the spore itself may become more porous. The increased porosity may allow the spore to adsorb some residual DPA that is present in the supernatant, thus resulting in pellet DPA that was greater at 500 μs than at pre-treatment. At greater treatment times, the holes caused by PEF treatment penetrate the spore coats, causing leakage of DPA from the spore core. Results indicate that PEF treatment at 2000 μs (E = 35 kV/cm) released 0.118 mM DPA per 10 mg dry weight spores, and decreased spore viability by 1.1 log. These data suggest that DPA is released from the spores, as a result of spore inactivation due to PEF treatment.

CONCLUSIONS

This study shows that pulsed electric fields are effective in inactivating bacterial spores. Factors that affect bacterial spore inactivation by PEF treatment are identified. Increasing duration time from 3 to 12 μs, and concomitantly decreasing frequency from 2000 to 500 Hz, does not affect survival rate of spores. There is an optimum PEF treatment temperature (30°C to 36°C) for inactivation of bacterial spores. L-Alanine significantly enhances the bacterial spore inactivation by PEF. The inactivation rate of bacterial spores increases when the electric field strength and total PEF treatment time increase.

SEM micrographs suggest that the mechanism of PEF inactivating bacterial spores is related to reversible and irreversible dielectric breakdown. The structures of PEF-treated bacterial spores are distinguishable from those of the untreated ones. The PEF-treated bacterial spores have similar structural changes when compared to the thermally inactivated bacterial spores. DPA release correlates with spore inactivation from PEF treatment. The detection of DPA in the supernatant would probably be a more convenient method for indicating spore inactivation than SEM or cell viability.

ACKNOWLEDGEMENT

This study was funded by the Ohio Agricultural Research and Development Center and the U.S. Army Natick Research Development and Engineering Center.

REFERENCES

Dunn, J. E. and Pearlman, J. S. 1987. Methods and apparatus for extending the shelf life of fluid food products. U.S. Patent, 4,695,472.

Gupta, R. P. and Murray, W. 1988. Pulsed high electric field sterilization. *IEEE Pulsed Power Conference Record*, pp. 58–64.

Hamilton, W. A. and Sale, A. J. H. 1967. Effects of high electric fields on microorganisms. II. Mechanism of action of the lethal effect. *Biochimica et Biophysica Acta,* 148:789–800.

Hulsheger, H. and Niemann, E.-G. 1980. Lethal effects of high voltage pulses on *E. coli* K 12. *Radiation and Enviromental Biophysics.* 18:281–288.

Hulsheger, H., Potel, J., and Niemann. E. -G. 1981. Killing of bacteria with electric pulses of high voltage strength. *Radiation and Enviromental Biophysics,* 20:53–65.

Janssen, F. W., Lund, A. J., and Anderson. L. E. 1958. Colorimetric assay for dipicolinic acid in bacterial spores. *Science,* 127:26–27.

Jayaram, S., Castle, G. S. P., and Margritis. A. 1992. Kinetics of sterilization of *Lactobacillus brevis* cells by the application of high voltage pulses. *Biotechnology and Bioengineering,* 40:1412–1420.

Jayaram, S., Castle, G. S. P., and Argyrios. M. 1993. The effects of high field DC pulse and liquid medium conductivity on surviviability of *Lactobacillus brevis. Applied Microbiology and Biotechnology,* 40:117–122.

Levinson, H. S. and Hyatt, M. T. 1970. Effects of temperature on activation, germination, and outgrowth of *Bacillus magaterium* spores. *Journal of Bacteriology,* 101:58–64.

Matsumoto, Y., Satake, T., Shioji, N., and Sakuma. A. 1991. Inactivation of microorganisms by pulsed high voltage applications. *Conference Record of IEEE Industrial Applications Society Annual Meeting,* pp. 652–659.

Mizuno, A. and Hori, Y. 1988. Destruction of living cells by pulsed high-voltage applications. *Transaction of IEEE Industrial Applications.* 24:387–394.

Palacios. P., Burgos, J., Hoz, L., Sanz, B.. and Ordoñez, J. A. 1991. Study of substances released by ultrasonic treatment form *Bacillus stearothermophilus* spores. *J. Appl. Bacteriol.* 71:445–451.

Rotman, Y. and Fields, M. L. 1968. A modified reagent for dipicolinic acid analysis. *Anal. Biochem.* 22:168.

Sale, A. J. H. and Hamilton, W. A. 1967. Effects of high electric fields on microorganisms. I. Killing of bacteria and yeasts, *Biochimica et Biophysica Acta,* 148:781–788.

Sale, A. J. H. and Hamilton, W. A. 1968. Effects of high electric fields on microorganisms. III. Lysis of erythrocytes and protoplasts. *Biochimica et Biophysica Acta,* 163:37–43.

Setlow, P. 1995. Mechanisms for the prevention of damage to DNA in spores of *Bacillus* species. *Annual Review of Microbiology,* 49:29–54.

Wax. R. and Freese, E. 1968. Initiation of the germination of *Bacillus subtilis* spore by a combination of compounds in place of L-alanine. *Journal of Bacteriology,* 95:433–438.

Wax, R., Freese, E., and Cashel, M. 1967. Separation of two functional roles of L-alanine in the initiation of *Bacillus subtilis* spore germination. *Journal of Bacteriology,* 94:522–529.

Yonemoto, Y., Yamashita, T., Muraji, M., Tatebe, W., Ooshima, H., Kato, J., Kimura, A., and Murata, K. 1993. Resistance of yeast and bacterial spores to high voltage electric pulses. *Journal of Fermentation and Bioengineering,* 75:99–102.

Zhang, Q., Chang, F. J., Barbosa-Cánovas, G.V., and Swanson, B. G. 1994a. Inactivation of microorganisms in semisolid foods using high voltage pulsed electric fields. *Food Science and Technology (lwt),* 27:538.

Zhang, Q., Monsalve-González, A., Qin, B., Barbosa-Cánovas, G. V., and Swanson, B. G. 1994b. Inactivation of *Saccharomyces cerevisiae* in apple juice by square wave and exponential decay pulsed electric fields. *Journal of Food Processing and Engineering,* 17:469–478.

Zhang, Q., Barbosa-Cánovas, G. V., and Swanson, B. G. 1995a. Engineering aspects of pulsed electric field pasteurization. *Journal of Food Engineering,* 25:261–281.

Zhang, Q., Qin, B., Barbosa-Cánovas, G. V., and Swanson, B.G. 1995b. Inactivation of *E. coli* for food pasteurization by high intensity short duration pulsed electric fields. *Journal of Food Process and Preservation,* 19:103–118.

Zimmermann, U. 1986. Electric breakdown, electropermeabilization and electrofusion. *Review of Physiology Biochemistry Pharmacology,* 105:175–256.

CHAPTER 12

Pulsed Electric Field and High Hydrostatic Pressure Induced Leakage of Cellular Material from *Saccharomyces cerevisiae*

S. L. HARRISON
G. V. BARBOSA-CÁNOVAS
B. G. SWANSON

ABSTRACT

H IGH intensity pulsed electric fields (PEF) and high hydrostatic pressure (HHP) are two innovative nonthermal processing techniques used to inactivate *Saccharomyces cerevisiae* in model and food systems. Membrane permeability is reported as the causative factor for inactivation of *Saccharomyces cerevisiae* treated by either PEF or HHP. Leakage of adenosine triphosphate (ATP), nucleic acid, and protein were studied for PEF and HHP treated *S. cerevisiae*. Relative amounts of ATP, nucleic acid, and protein released from PEF and HHP treated *S. cerevisiae* did not increase proportionately to increased number of pulses with PEF treatment, or for increased pressure with HHP treatment.

INTRODUCTION

High intensity pulsed electric field (PEF) and high hydrostatic pressure (HHP) treatments are two innovative nonthermal food processing methods being investigated by food scientists. Both PEF and HHP technologies yield food products of higher quality with respect to nutrients and flavor as compared to heat treated counterparts (Jayaram et al., 1992; Kimura et al., 1994). Many of the microbiological inactivation limitations for PEF and HHP techniques are determined for currently available equipment (Zhang et al., 1995; Hülsheger et al., 1981; Palaniappan et al., 1990). However, the mechanism of microbiological inactivation is still uncertain. Elucidation of the mechanism(s) responsible for microbial inactivation will expedite the introduction of the next generation of PEF and HHP food processing equipment.

Membrane permeability is the most widely accepted theory describing the inactivation effects of PEF (Castro et al., 1993) and HHP (Morita, 1975). Scanning electron microscopy (SEM) and transmission electron microscopy (TEM) are tools used to study morphological changes to *Saccharomyces cerevisiae*. Using SEM and TEM techniques, observed changes to the cell walls of *S. cerevisiae* treated by PEF (suggesting the possibility of leakage of cellular contents) include pinholes (Hayamizu et al., 1989), craters or holes (Mizuno and Hayamizu, 1989), and voids (Harrison et al., 1997). Similarly, SEM and TEM structural studies of *S. cerevisiae* treated by HHP were conducted by Shimada et al. (1993). In the original PEF studies by Sale and Hamilton (1967), no evidence was obtained for the destruction of the characteristic bimolecular structure of membranes by PEF treatment, as observed with electron microscopy. Grahl et al. (1992) stated that SEM did not show evidence of destruction or damage to the surface of *S. cerevisiae* after PEF treatment. A correlation between observed cell wall damage and *S. cerevisiae* inactivation by either PEF or HHP has not been reported.

Disruption of cellular organelles, specifically loss of ribosome bodies, is an alternative theory, termed organelle disruption, for *S. cerevisiae* inactivation with PEF (Harrison et al., 1997). Similar destruction to *S. cerevisiae* is reported for HHP treatments (Amesz et al., 1973; Smith et al., 1975; Shimada et al., 1993). Harrison et al., (1997) reported a 99.99% reduction in S. cerevisiae viability, with a total loss of ribosome bodies, after PEF treatment at 40 kV/cm and 64 pulses. Conversely, only 0.1% of the yeast cells observed by TEM techniques demonstrated electroporation damage.

Leakage of cytosolic material will accompany formation of cell wall pores. Measuring substances normally inside the yeast cell, but not normally present outside the cell, serves as a good indicator of cell wall pore formation and subsequent leakage of cellular material. As the pore size increases, larger molecules will pass through the membrane and outside of the cell. Adenosine triphosphate (ATP) is a small molecule found almost completely contained within *S. cerevisiae* cells. Fluorescence assays utilizing the luciferase/D-luciferin enzyme system are used to quantify microbial ATP (Stanley, 1989). Larger molecules are also of interest. A quick estimate of both the protein and the nucleic acid content can be made by measuring UV light absorption at 280 and 260 nm (McGilvery, 1970). The release of 260 nm absorbing material from PEF treated *E. coli* 8196 (Sale and Hamilton, 1967) and *S. cerevisiae* (Grahl et al., 1992) is reported. The objective of the current research was to treat *Saccharomyces cerevisiae* with PEF and HHP, and correlate *S. cerevisiae* inactivation with leakage of cellular ATP, nucleic acid, and protein.

MATERIALS AND METHODS

Each experiment was repeated twice. Each treatment within an experiment was run in triplicate. One set of PEF experiments and one set of HHP experiments were run on day one. The PEF and HHP experiments were repeated on day two with freshly prepared chemicals.

SACCHAROMYCES CEREVISIAE INACTIVATION

Saccharomyces cerevisiae (ATCC 16664) were cultured in yeast malt broth (DIFCO 0711-01-9), contained in Erlenmeyer flasks, with continuous agitation, in a temperature controlled shaker (model MSB-3322A-I, GS Blue Electric, Blue Island, IL), at 24°C, until reaching an absorbance of 210 Klett units (early stationary phase), with a viable count of 7.1×10^7 cfu/mL. Erlenmeyer flasks containing the yeast cells were placed in an ice bath for 5 min, transferred to 15 mL sterile disposable centrifuge tubes, pelleted for 15 min at 4000 g and 5°C in a Beckman (model J2-HS, Beckman Instrument Inc., Palo Alto, CA) centrifuge. The supernatant was discarded, the pellet resuspended, washed with an equal volume of 5°C nutrient broth, and centrifuged. The resuspension and centrifugation procedure was repeated twice. The resultant pellet ($\sim$1 g) was suspended in 1 mL of 20% (v/v) glycerol and frozen at −70°C until needed.

Individual frozen yeast/glycerol pellets were centrifuged in a Sorvall (model RT 6000B, DuPont Company, Newtown, CT) centrifuge for 1 min at 3390 relative centrifugal force (RCF) and 5°C. The glycerol supernatant was removed. Each yeast pellet was resuspended in 5 mL, pH 7.2, 10 mM Tris(hydroxymethyl) aminomethane buffer containing 10 mM $MgSO_4$ (Tris buffer), previously treated with 0.05 U/mL, grade VIII Apyrase (an ATPase) from potato (Sigma Chemical Company, St. Louis, MO), for 16 hr, followed by autoclaving. Apyrase enzyme was added to the Tris buffer to remove contaminating ATP (Lyman and DeVincenzo, 1967). Subsequent autoclaving was performed to inactivate the apyrase enzyme. Non-disposable glassware was also treated overnight with apyrase followed by autoclaving. The yeast/Tris buffer solution was centrifuged at 3390 RCF for 1 min, and the supernatant was removed and saved for analysis. The yeast pellet was resuspended in a second 5 mL aliquot of Tris buffer, followed by centrifugation at 3390 RCF for 2 min. The supernatant was removed and saved for analysis. Yeast pellets were suspended at a ratio of one pellet/100 mL in Tris buffer, and equilibrated at room temperature ($\sim$22°C) for thirty min prior to PEF or HHP treatment.

PULSED ELECTRIC FIELD TREATMENT

A pilot plant size pulser (Physics International, San Leandro, CA), in conjunction with an autoclaved sterile recirculating coaxial electrode chamber (Zhang et al., 1995), was utilized for PEF treatments. Observable gas bubbles were removed from the treatment chamber after addition of yeast/Tris buffer solution, and prior to application of PEF, to reduce the chance of arcing. The PEF operation was conducted at room temperature ($\sim$22°C). An electric field of 65 kV/cm, capacitance of 0.5 μF, and exponential decay pulses of 4 μs were utilized. The yeast/Tris buffer solution was sampled prior to treatment and after 1, 2, 4, 6, 8, and 10 pulses. The chamber was rinsed with 70% ethanol, followed by a sterile deionized water rinse between experiments.

HIGH HYDROSTATIC PRESSURE TREATMENT

A pilot plant scale hydrostatic pressure device (Engineered Pressure Systems Inc., Andover, MA) was utilized for HHP treatments. Food grade 4 mil polyethylene pouches (Power Plastics Inc., Paterson, NJ) were filled with 20 mL of yeast/Tris buffer solution and heat sealed after removing as much air as possible. HHP solutions were pressurized to 20, 40, 60, 80, or 100 kpsi and held for 1 min. The come-up pressure was a function of time, and approximated 25 kpsi/min.

SACCHAROMYCES CEREVISIAE VIABILITY

The viability of *S. cerevisiae* before and after PEF and HHP treatments was monitored by counting colony forming units (cfu) in potato dextrose agar (PDA) (DIFCO 0012-01-5) plates. The PEF and HHP treated yeast/Tris buffer solutions were serially diluted with 0.1% sterile peptone (DIFCO 01118-01-8) solution, plated in PDA acidified with 14 mL/L of filter sterilized 10% tartaric acid solution. The PDA plates were incubated at 24°C for 72 hr to obtain *S. cerevisiae* viable counts. Serial dilutions for the viable counts were performed to obtain cfu in the agar plate of 25 to 300. The mean viable count was calculated from four plates per treatment.

PREPARATION OF SAMPLES AND CONTROLS FOR ANALYSIS

Five milliliters of untreated yeast/Tris buffer solution served as the negative control. The positive control consisted of 5 mL untreated yeast/Tris buffer solution in a test tube placed in boiling water for 5 min. After heating, the negative control was treated the same as the positive control, PEF treated solutions, and HHP treated solutions. Aliquots of 5 mL each of treated or control solutions were placed in 15 mL conical sterile plastic centrifuge tubes. Controls, PEF treated, and HHP treated solutions were centrifuged in a Sorvall (model RT 6000B, DuPont Company, Newtown, CT) centrifuge at 3390 RCF for 1 min, the supernatant was removed and saved for analysis.

ATP DETERMINATION

An Aminco-Bowman spectrophotofluorometer (model J4-8961, American Instrument Company, Silver Spring, MD) was utilized for fluorescence determination of ATP. Luciferase from *P. fischeri* and synthetic D-luciferin were obtained from ICN Pharmaceutical Inc. (Costa Mesa, CA). Other chemicals were reagent grade. Solutions were made up in the Tris buffer described previously. A 350 μL aliquot of control, PEF treated, or HHP treated sample was placed in a disposable cuvette. A 50 μL aliquot of 10 mM luciferin was added and mixed. A 50 μL aliquot of 5.0 mg/mL luciferase was added, and the fluorescence at 562 nm was determined within 5 s.

PROTEIN AND NUCLEIC ACID DETERMINATION

A Perkin-Elmer UV/VIS spectrophotometer (model Lambda 2, Funkentstört, Germany) was utilized for determination of nucleic acid and protein leakage from PEF and HHP treated *S. cerevisiae*. A 2 mL aliquot of control, PEF treated, or HHP treated supernatant was placed in a quartz cuvette. Relative nucleic acid and protein leakage was determined by measuring the absorbance of the supernatant at 260 and 280 nm. The Tris buffer served as the blank. Significant differences between means were established at $P \leq 0.05$, and determined using least square means (SAS, 1998).

RESULTS AND DISCUSSION

The Aminco-Bowman spectrophotofluorometer used for ATP determinations lacked the sensitivity to sufficiently discriminate relative ATP leakage from *S. cerevisiae*. Adjusting the spectrophotofluorometer to maximum sensitivity was required to achieve a measurable signal. Statistical differences in ATP leakage from *S. cerevisiae* were not observed as the number of pulses increased for PEF treatments, or as the pressure increased for HHP treatments.

PULSED ELECTRIC FIELD TREATMENT

As much as a one log reduction in *S. cerevisiae* viability may occur during the freezing step associated with storing *S. cerevisiae* at −70°C, and the subsequent thawing of the yeast cells prior to an experiment. Freeze-thaw inactivation

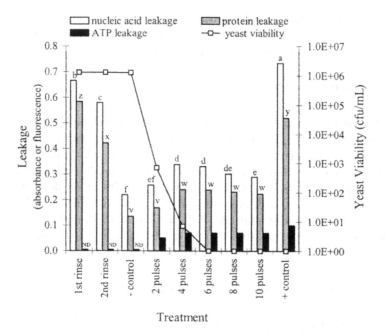

Figure 12.1 *Saccharomyces cerevisiae* inactivation and leakage of cellular material after 65 kV/cm pulse electric field treatment. Bars with the same letters are not significantly different ($P \leq 0.05$). Letters a–f are not intended to be compared with letters v–z. Letters a–f correspond to nucleic acid leakage. Letters v–z correspond to protein leakage. ND = not detected.

of *S. cerevisiae* resulting in membrane and cell wall damage is indicated in Figure 12.1, with the nucleic acid and protein leakage observed for the first and second rinsing of the yeast cells with Tris buffer. The *S. cerevisiae,* damaged during freezing and thawing, contributed to the relatively high amounts of nucleic acid and protein observed in the two Tris buffer rinses. No ATP was detected in the two Tris buffer rinse solutions, or in the control. Enzymatic activity persisted after disruption of the yeast cells, resulting in the depletion of free ATP. Compared to the negative control, the amount of material released from the *S. cerevisiae* was expected to rise as the number of applied pulses increased. An increase in ATP, nucleic acid and protein was observed as the number of PEF pulses increased from zero to 4 pulses (Figure 12.1). From pulse number 4 to 10 pulses, the amount of ATP leakage remained unchanged. The observed nucleic acid and protein leakage actually decreased in the range of 4 to 10 pulses. Sale and Hamilton (1967) also reported a moderate increase, followed by a slight decrease, in 260 nm absorbing material for *E. coli* treated by PEF. The maximum release of 260 nm absorbing material occurred after 4 pulses, and corresponds to approximately half of the value observed for the heated positive control. Grahl et al. (1992) reported that PEF treatment of

S. cerevisiae only caused the release of 50% of the RNA and ribonucleotides as compared to the amount released for the ultrasonic treated positive control.

If the primary mechanism of *S. cerevisiae* inactivation is membrane permeability, then a sharp increase in cytosolic leakage will accompany *S. cerevisiae* inactivation. As depicted in Figure 12.1, a five log reduction in yeast viability took place between zero and 4 pulses. Conversely, only a gradual rise in ATP, nucleic acid, and protein leakage takes place over this range. Of even greater interest is the observation that the leakage of the three cellular components, monitored after PEF treatment of *S. cerevisiae,* remained the same or decreased for PEF treated solutions with greater than 4 pulses. The lack of correlation between yeast inactivation and leakage, in conjunction with a reported correlation between *S. cerevisiae* inactivation and cellular organelle disruption (Harrison et al., 1997), supports the organelle disruption mechanism theory.

HIGH HYDROSTATIC PRESSURE TREATMENT

Analysis of the first and second Tris buffer rinse solutions and the negative control in this set of experiments parallels the results of the PEF experiments. The ATP, nucleic acid, and protein leakage data obtained for ATP leakage from HHP treated *S. cerevisiae* cells (Figure 12.2) were similar to the results obtained

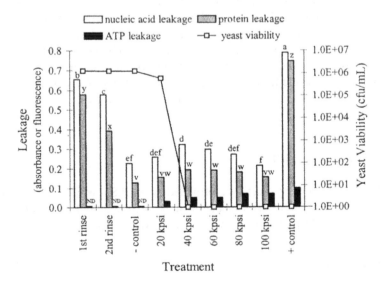

Figure 12.2 *Saccharomyces cerevisiae* inactivation and leakage of cellular material after high hydrostatic pressure treatment with 1 min at desired pressure. Bars with the same letters are not significantly different ($P \leq 0.05$). Letters a–f are not intended to be compared with letters v–z. Letters a–f correspond to nucleic acid leakage. Letters v–z correspond to protein leakage. ND = not detected.

from the PEF experiments (Figure 12.1). As the HHP treatment pressure increased from zero to 40 kpsi, the amount of nucleic acid and protein released from the yeast cells increased. From 40 kpsi to 100 kpsi, the amount of nucleic acid and protein leakage decreased. Similar research conducted with *Saccharomyces cerevisiae* by Shimada et al. (1993) yielded data showing a continuous increase in material absorbing at 260 nm at HHP pressures of 29 kpsi to 73 kpsi. Decreasing nucleic acid and protein content observed in the absorbance assays, after PEF and HHP treatment, may be attributed to precipitation reactions between protein and nucleic acid constituents contained in the assay solution. Organelle disruption also results from HHP processing of *Saccharomyces cerevisiae* (Shimada et al., 1993; Bang, 1996).

As the viability of *S. cerevisiae* decreases from 10^6 yeast cells/mL at atmospheric pressure, to zero viable yeast cells after 1 min at 40 kpsi, leakage of nucleic acid and protein from HHP treated *S. cerevisiae* increases (Figure 12.2). At pressures greater than 40 kpsi the nucleic acid and protein leakage decrease.

CONCLUSIONS

Treatment of *Saccharomyces cerevisiae* by either pulsed electric field or high hydrostatic pressure results in leakage of ATP, nucleic acid, and proteins out of the cell. A definitive correlation does not exist between *S. cerevisiae* inactivation and release of nucleic acid, protein, or ATP. Transmission electron microscopy observations suggest organelle disruption as the primary mechanism involved in yeast inactivation for both PEF and HHP treatments. The PEF and HHP organelle disruption evidence, along with the lack of correlation between inactivation and leakage of cellular material from PEF and HHP treated *S. cerevisiae*, suggests the need for further scrutiny of the PEF membrane permeability theory.

ACKNOWLEDGEMENTS

The funding for this research was provided by the U.S. Army Natick Research Development and Engineering Center, Natick, MA, and the Bonneville Power Administration, Department of Energy, Walla Walla, WA.

REFERENCES

Amesz, W. J. C., Bevers, E. M., and Bloemers, H. P. J. 1973. Dissociation of yeast ribosomes, the effects of hydrostatic pressure and KCl. *Molecular Biology Reports* 1, 33–39.

Bang, W. S. 1996. Unpublished data. Department of Food Science and Human Nutrition, Washington State University, Pullman, WA.

Castro, A. J., Barbosa-Cánovas, G. V., and Swanson, B. G. 1993. Microbial inactivation of foods by pulsed electric fields. *J. Food Processing Preservation* 17, 47–73.

Grahl, T., Sitzmann, W., and Märkl, H. 1992. Killing of microorganisms in fluid media by high-voltage pulses. *DECHEMA Biotechnology Conference Series* 5B, 675–678.

Harrison, S. L., Barbosa-Cánovas, G. V., and Swanson, B. G. 1997. *Saccharomyces cerevisiae* structural changes induced by pulsed electric field treatment. *Lebensm.-Wiss. U.-Technol.* 30(3), 236–240.

Hayamizu, M., Tenma, T., and Mizuno, A. 1989. Destruction of yeast cells by pulsed high voltage application. *J. Institute Electrostatics Japan* 13, 322–331.

Hülsheger, H., Potel, J., and Niemann, E.-G. 1981. Killing of bacteria with electric pulses of high field strength. *Radiation and Environ. Biophys.* 20, 53–65.

Jayaram, S., Castle, G. S. P., and Margaritis, A. 1992. Kinetics of sterilization of *Lactobacillus brevis* cells by the application of high voltage pulses. *Biotech. Bioeng.* 40, 1412–1420.

Kimura, K., Ida, M., Yosida, Y., Ohki, K., Fukumoto, T., and Sakui, N. 1994. Comparison of keeping quality between pressure-processed jam and heat-processed jam: Changes in flavor components, hue, and nutrients during storage. *Biosci. Biotech. Biochem.* 58, 1386–1391.

Lyman, G. E. and DeVincenzo, J. P. 1967. Determination of picogram amounts of ATP using the luciferin-luciferase enzyme system. *Anal. Biochem.* 21, 435–443.

McGilvery, R. W. (Ed.). 1970. *Biochemistry: A Functional Approach*, 1st ed. Saunders Co., Philadelphia, PA.

Mizuno, A. and Hayamizu, M. 1989. Destruction of bacteria by pulsed high voltage application. *Sixth International Symposium on High Voltage Engineering*, New Orleans, LA, USA. Aug. 28–Sept. 1, 1989.

Morita, R. Y. 1975. Psychrophilic bacteria. *Bacteriol. Rev.* 39(2), 144.

Palaniappan, S., Sastry, S. K., and Richter, E. R. 1990. Effects of electricity on microorganisms: A review. *J. Food Proc. Pres.* 14, 393–414.

Sale, A. J. H. and Hamilton, W. A. 1967. Effects of high electric fields on microorganisms: I. Killing of bacteria and yeast. *Biochim. Biophys. Acta.* 148, 781–788.

SAS. 1998. *SAS User's Guide.* Statistical Analysis Systems Institute, Cary, NC.

Shimada, S., Andou, M., Naito, N., Yamada, N., Osumi, M., and Hayashi, R. 1993. Effects of hydrostatic pressure on the ultrastructure and leakage of internal substances in the yeast *Saccharomyces cerevisiae. Appl. Microbiol. Biotechnol.* 40, 123–131.

Smith, W., Pope, D., and Landau, J. V. 1975. Role of bacterial ribosome subunits in barotolerance. *J. Bact.* 124, 582–584.

Stanley, P. E. 1989. A concise beginner's guide to rapid microbiology using adenosine triphosphate (ATP) and luminescence. Chapter 1 in *ATP Luminescence Rapid Methods in Microbiology*, P. E. Stanley, B. J. McCarthy, and R. Smither (Eds.), pp. 1–10. Blackwell Scientific Publications, Boston, MA.

Zhang, Q., Barbosa-Cánovas, G. V., and Swanson, B. G. 1995. Engineering aspects of pulsed electric field pasteurization. *J. Food Eng.* 25, 261–281.

CHAPTER 13

Nonthermal Inactivation of *Pseudomonas fluorescens* in Liquid Whole Egg

M. M. GÓNGORA-NIETO
L. SEIGNOUR
P. RIQUET
P. M. DAVIDSON
G. V. BARBOSA-CÁNOVAS
B. G. SWANSON

ABSTRACT

THE bacteria, *Pseudomonas fluorescens,* can grow during refrigerated storage, causing storage problems and thus becoming a threat to the quality and shelf life of liquid whole egg (LWE) products. This bacterium was the principal component isolated from raw and spoiled LWE pretreated by pulsed electric fields (PEF). Heat pasteurization is effective for the inactivation of these types of bacterial flora, but due to the high protein content of LWE, some protein coagulation may occur. This problem might be overcome by nonthermal treatments.

The objective of this study was to evaluate the inactivation of different strains of *P. fluorescens* in LWE by a hurdle approach of pulsed electric fields (PEF) or high hydrostatic pressure (HHP) and antimicrobials.

LWE, inoculated with WSU-07, ATCC 17400 or ATCC 13525 strains of *P. fluorescens,* was processed by PEF (117 pulses, 48 kV/cm) or HHP [5 min at 20–40 kpsi (133.33 Mpa–266.6 Mpas)], alone or in combination with the following antimicrobials: parabens mixture (0–0.1%), citric acid (0.15–0.5%), parabens with citric acid, parabens with EDTA (0.1%:5 mM), nisin (200 UI), nisin with EDTA (200 UI:5 mM), lysozyme (0–1000 UI), and lysozyme with EDTA (500 UI:5 mM). Appearance changes after treatments were monitored by color measurement. Ultrastuctural damage was evaluated by electron microscopy techniques.

A synergistic behavior (>90% inactivation enhancement) was found between the nonthermal technologies and citric acid (CA), citric acid in combination with parabens, nisin, and lysozyme. A significant difference in the inactivation rate was detected among the different *P. fluorescens* strains. The WSU-07 strain was significantly more resistant to PEF, alone and in combination with antimicrobials, than the other two ATCC strains. The inactivation kinetic constants were evaluated for all the strains inactivated by PEF. Pulsed electric field technology reduced the population of *P. fluorescens* suspended in LWE by more than 4 log cycles in 0.00023 seconds of treatment time under an electric field of 48 kV/cm; an equivalent microbial reduction was achieved with 5 min HHP at 30 kpsi. HHP treatments resulted in a slight change in color for pressures higher than 30 kpsi, while PEF had no effect at up to 117 pulses. Ultrastructural damage evaluation evidenced damage of the cells membrane.

INTRODUCTION

Processed liquid egg products are those obtained after breaking and separating shell eggs, followed by pasteurizing and packing. Processed liquid egg products are convenient for commercial baking, food service, and household use. Egg products are frequently preferred to shell eggs because they have many advantages, including expediency, labor saving, minimal storage and waste disposal requirements, ease of portion control, and product quality, stability and uniformity. In the United States, egg production has steadily increased in the past decade (National Agricultural Statistics Service, 1999a), with a minimum growth of 10.8% from November 1993 (71.9 billion eggs) to November 1998 (79.7 billion eggs). Furthermore, broken shell eggs, based on a census of all commercial egg breaking and processing plants, increased 11% from November 1996 (124 million dozens) to November 1998 (133 million dozens) (National Agricultural Statistics Service, 1999b).

At present, pasteurization of all egg products is mandatory; therefore, liquid egg producers must provide consumers with a safe product that is comparable in flavor, nutritional value, and most functional properties to shell eggs. Today's industry pasteurizes liquid egg products with a heat treatment of 64.4°C for 2 to 5 min to ensure the destruction of *Salmonella* spp. (Delves-Broughton et al., 1992), to minimize the presence of spoilage microorganisms such as *Pseudomonas fluorescens,* and to meet other bacteriological standards (coliforms, yeast, and molds counts). While such a thermal process inactivates undesirable microorganisms, it also affects the quality attributes of the treated food (e.g., flavor, color, aroma, nutrients, functional properties), and is high in energy demand (Barbosa-Cánovas et al., 1999). For these reasons nonthermal preservation processes are under intense research to evaluate their potential as alternative or complementary processes to traditional methods of food preservation. Such nonthermal technologies include high hydrostatic pressure (HHP), oscillating magnetic fields, high intensity pulsed electric fields (PEF), intense light pulses, irradiation, and antimicrobials.

Both PEF and HHP have been shown to be effective for the inactivation of gram-positive bacteria, gram-negative bacteria, yeasts, and molds (Martín-Belloso et al., 1997; Ho et al., 1995; Qin et al., 1998; Barbosa-Cánovas et al., 1998, 1999). The factors affecting microbial inactivation by PEF are electric field strength, treatment time (number of pulses times the pulse duration), pulse shape, temperature of the medium, growth stage of the microbiological flora, and electrical properties of the food to be treated (Qin et al., 1996). In HHP treatment, the microbial inactivation depends on magnitude and duration of the high pressure treatment, type and number of microorganisms, temperature, and composition of the suspension medium or food (Palou, 1998).

The use and action of chemical antimicrobials is well characterized in microbiological media (Cutter and Siragusa, 1995; Hauben et al., 1996), but is not well documented in complex foods like LWE. A majority of antimicrobials are organic acids. They are generally most active against microorganisms in the undissociated form. Therefore, selection of an antimicrobial must take into account its pK_a, which should be higher than the pH of the food (Davidson, 1992). In addition, there are many other factors that must be considered, such as the type of microorganism (e.g., gram-positive bacteria, gram-negative bacteria, yeast, mold, spore), mode of action (e.g., cell wall degradation, chelation), physical and chemical properties of the antimicrobial (e.g., hydrophylic, hydrophobic, etc.), the type and intensity of any further treatment (e.g., heat treatment, other nonthermal processes), and the type of food and its physical and chemical properties (e.g., pH, fat content, presence of ions, water activity).

The pH of LWE is around 7.5. In this range, parabens are one of the few potentially active antimicrobials because they remain undissociated up to a pH of 8.5. In the United States, methyl and propyl parabens are generally recognized as safe (GRAS) at a maximum concentration of 0.1% each. When used in combination, the total may not exceed 0.1% (Code of Federal Regulations, 1997).

To preserve the bright yellow color of liquid whole egg during refrigerated storage, it is necessary to add citric acid (0.1 to 0.5%). The presence of citric acid can also have antimicrobial properties: an increase in the food acidity creates an unfavorable environment that limits the growth of the microorganisms (Doores, 1992). An increase in acidity will also lower the pH, and may enhance the effectiveness of other antimicrobials. It has been shown that a wide range of antimicrobials have no effect against *Pseudomonas* (Branen and Davidson, 1998). Therefore, an attractive alternative is the use of multiple barriers against the microorganisms, well known as the "hurdle" technology approach (Hauben et al., 1996; Liu et al., 1996).

The objective of the present study was to inactivate *Pseudomonas fluorescens* inoculated in LWE by nonthermal processes (high hydrostatic pressure and pulsed electric fields), in combination with chemical (parabens and citric acid) and natural (nisin and lysozyme) antimicrobials (alone or in combination with EDTA as a chelating agent).

MATERIALS AND METHODS

PREPARATION OF LIQUID WHOLE EGG

Fresh eggs from a local supermarket were inspected for integrity of the shell, hand washed (twice) with 200 ppm chlorine solution, and air dried before breaking the shell under aseptic conditions. The contents of selected eggs were removed and beaten using a sterile Hobart mixer (Hobart Manufacturing, Inc., Troy, OH) for 5 min at speed "1." The homogeneous liquid whole egg was filtered three times with a sterile sieve (kitchen sifter), and stored in sterile glass containers until use.

PSEUDOMONAS STRAINS AND PREPARATION OF INOCULUM

Three strains of *Pseudomonas fluorescens* were used. The first was an ATCC 17400 biotype C culture, isolated from eggs, and the second was a *P. fluorescens* ATCC 13525. Both were obtained from the American Type Culture Collection (ATCC). The cultures were reactivated as indicated by the ATCC manual (ATCC, 2000). The third strain was a pure culture (WSU-07) isolated from spoiled liquid whole eggs that had been pasteurized by PEF, from the collection of the Center for Nonthermal Processing of Food (Washington State University, Pullman, WA). This WSU-07 strain was isolated during the conduction of shelf life studies of PEF treated LWE. After 21 days of refrigerated storage at $4°C$, the aerobic plate count increased above regulatory limits (20,000 to 25,000 cfu/mL), determining the end of the LWE shelf life. The same *P. fluorescens* was isolated from plate count agars ($35°C$ 48 h) on 7 different spoiled egg samples (from different PEF batches).

The three strains were maintained on nutrient agar slants (Difco, Detroit, MI), refrigerated, and transferred weekly to assure activity. Prior to PEF testing, a loop of the cultures from the refrigerated slants was incubated in 50 mL or 100 mL of tryptic soy broth (Difco, Detroit, MI) for 15 hr at $32°C$, with aeration from a shaker (166 RPM, GS Blue Electric, Blue Island, IL). The $32°C$ incubation temperature was selected near the PEF processing temperature, and as an intermediate point between the optimum growth temperature for *P. fluorescens* spp. and that at which the WSU-07 strain was isolated. The end point of the incubation was when the cells were in stationary phase ($\sim N_0$, 10^7 cfu/mL). Cells were enumerated by plating in tryptic soy agar (TSA, Difco, Detroit, MI), using serial dilution in sterile 0.1% peptone, and incubated for

48 h at 32°C. Each liter of LWE was inoculated with 50 mL of inoculum, and stirred for 5 min before any treatment.

PREPARATION AND USE OF ANTIMICROBIALS

A mixture of methyl:propyl (3:1) parabens (*p*-hydroxy-benzoic acid methyl ester and *n* propyl *p*-hydroxy benzoate, Sigma, St. Louis, MO) was added to the LWE to achieve a final concentration of 0.05% or 0.1%. Parabens were dissolved in 40 mL of LWE just before the experiment, added to the inoculated LWE, and stirred for five minutes before PEF or HHP treatments. Citric acid (citric acid monohydrate, Sigma, St. Louis, MO) was dissolved in distilled water to prepare a 60% (w/v) stock solution that was autoclaved. The stock solution was added to the inoculated LWE to a final concentration of 0.15% or 0.5%. EDTA (ethylenediamine-tetraacetic acid, Fisher Scientific, Fair Lawn, NJ) was dissolved in distilled water and autoclaved. The 500 mM EDTA stock solution was added to a final concentration of 5 nM. Nisin [nisaplin, 2.5% nisin, 10^6 UI/mL (Aplin and Barret, Trowbridge, England)] was dissolved in sterile 0.02 N HCl (stock solution of 0.2 g nisaplin/mL, or 5 mg nisin/mL). The nisin stock solution was added to a final concentration of 200 mg of nisaplin per liter of LWE (5 mg of nisin/L of LWE, 5 µg/mL, 200,000 UI nisin/L or 200 UI/mL), for HHP treatments and from 0 to 1000 mg per liter of LWE for PEF treatments. Lysozyme (Sigma, St. Louis, MO) was dissolved, just before use, in distilled water to prepare a stock solution with a concentration of 40 mg/mL, and kept in ice until use. The final concentration of lysozyme in the egg was 500 µg/mL.

NONTHERMAL TREATMENTS

The study of microbial inactivation was performed independently by two nonthermal technologies, HHP and PEF treatments, in combination with the selected antimicrobials.

PULSED ELECTRIC FIELD TREATMENTS

A continuous treatment chamber, as described in Qin et al. (1995), with coaxial stainless steel electrodes, 28 mL capacity, and 0.6 cm gap, was used to apply high intensity pulsed electric field (PEF) treatments. The PEF processing conditions (Table 13.1) remained the same for all experiments, with a constant flow rate of 650 mL/min using a peristaltic pump (Masterflex Model 7564-00, Cole Parmer Instrument Co., Chicago, IL). The electric field was generated using a pilot plant size pulser manufactured by Physics International (San Leandro, CA). Such equipment allows the delivery of slightly underdamped pulses of 2 µs, with a pulsing rate set at 3 Hz, 40 kV of input voltage, and a 0.5 µF

TABLE 13.1. Treatment Conditions for Processing Liquid Egg by PEF.

Parameters	Operation Conditions
Capacitance (μF)	0.5
Input voltage (kV)	40
Input flow rate (L/min)	0.65
Input pulse rate (Hz)	3
Peak voltage (kV)	29
Electric field intensity (kV/cm)	48
Pulse energy (J)*	210
Maximum temperature (°C)	32

*Assuming the capacitor was charged at 29 kV.

capacitor. The electric field intensity of 48 kV/cm delivered to the chamber was determined with an oscilloscope (Hewlett Packard model 54520A, Colorado Springs, CO) and Equation (1).

$$E_{avg} = \left(\frac{2}{R_{hv} R_{lv}} \right) \left(\frac{V_0}{\ln(R_{lv}/R_{hv})} \right) \tag{1}$$

where R_{hv} and R_{lv} are the radii of the high and low voltage electrodes, respectively, and V_0 is the delivered voltage to the PEF chamber determined with the oscilloscope.

Under the specified processing conditions the PEF processed LWE received 7.7 pulses per unit volume in the treatment chamber. The temperature of the treated egg was measured at the inlet and exit of the treatment chamber using digital thermometers (John Fluke Mfg. Co., Everett, WA).

Gas bubbles were removed from the treatment chamber after addition of LWE and prior to PEF processing to reduce the possibility of arcing. A continuous recirculation treatment (Figure 13.1) over 35 min was conducted with 1.5 L of LWE in the system, applying a total of 117 pulses. The controlled LWE temperature was kept below 32°C. Samples were taken every 5 min or 17 pulses, and collected in sterile tubes that were cooled on ice until microbiological analysis. All experiments were conducted in duplicate and the microbial counts had 3 replicates.

HIGH HYDROSTATIC PRESSURE TREATMENTS

Triplicate 20 mL samples of inoculated LWE were placed in food grade 4 mil polyethylene pouches (Power Plastics Inc., Paterson, NJ). These bags were put into a second identical polyethylene pouch, and sterile distilled water was added to the first pouch, followed by removal of as much air as possible. The first pouch was then heat sealed. A pilot plant scale hydrostatic pressure system

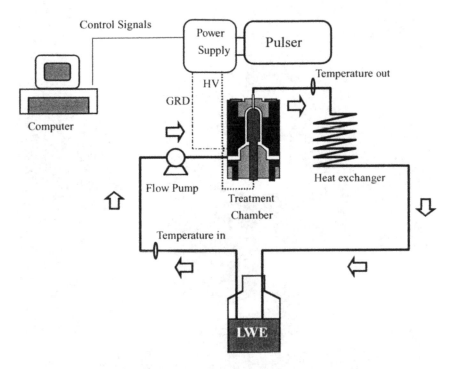

Figure 13.1 Continuous circulation pulsed electric field operation.

(Figure 13.2) (Engineered Pressure Systems, Inc., Andover, MA), operated at room temperature ($\sim$22°C), was utilized to process the LWE. The evaluated pressure ranged from 20 to 40 kpsi, with a come-up time from 120 to 180 sec (2 to 3 min), a holding time of 5 min, and a pressure release time of less than 15 s (Figure 13.3). All treatments were conducted in duplicate.

VIABILITY OF MICROORGANISMS

The viable number of *P. fluorescens* in LWE was determined prior to processing and immediately after processing. For enumeration, LWE was diluted in sterile 0.1% peptone (Difco, Detroit, MI), serially diluted, and pour plated using TSA. Pour plating in violet red bile agar (VRB, BBL, Cockeysville, MD) and TSA of the raw LWE was performed to verify the asepsis of the eggs and the egg breaking process. The *Compendium of Methods for the Microbiological Examination of Foods* (Vanderzant and Splittstoesser, 1992) was used as a guide for conducting microbial evaluation. For the experiments in which antimicrobials were added to the LWE, samples were plated without antimicrobials in the media.

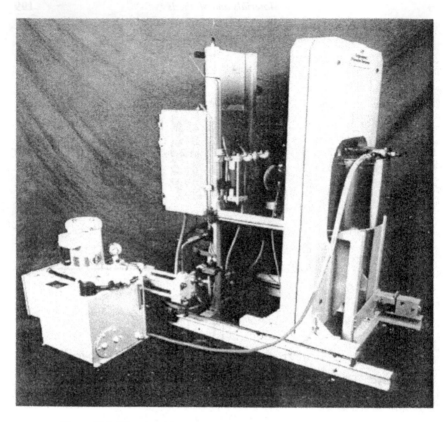

Figure 13.2 Pilot plant high hydrostatic pressure unit for batch processing.

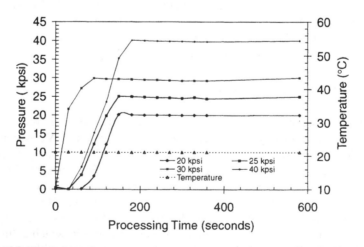

Figure 13.3 High hydrostatic pressure cycles (temperature and pressure profiles) for 20 to 40 kpsi.

200

EVIDENCE OF ULTRASTRUCTURAL DAMAGE
BY ELECTRON MICROSCOPY

Samples of stationary phase cells of *P. fluorescens* strains suspended in tryptic soy broth (TSB) and LWE, prior to treatment and after treatment, were fixed and processed according to the recommended procedure for preparing specimens for SEM and TEM. The samples were fixed in glutaraldehyde and paraformaldehyde (stock solution: 0.2 M PIPES buffer containing 2% glutaraldehyde and 2% paraformaldehyde), and held in refrigerated storage for 24 h. Those samples suspended in TSB were turned into pellets, while those suspended in LWE were cut into 1 mm cubes (the egg samples formed gels with the glutaraldehyde and paraformaldchyde). All samples were rinsed with 0.1 M of phosphate buffer (pH 7.2), and post-fixed in 1% osmium tetroxide for 12 h. The fixation was followed by rinsing twice with 0.1 M phosphate buffer (pH 7.2). For the rinsing step, the supernatant was discarded by centrifugation at 15,850 · gravity in a Beckman model Microfuge E (Palo Alto, CA), and the pellets were rinsed twice for 10 min each time. The samples were split in two halves, one for the SEM analysis, and the other for the TEM analysis.

For TEM, the cells were sequentially dehydrated with 30, 50, 70, 95, and 100% acetone for 10 min each, and infiltrated and embedded in Spurr's resin, starting with a concentration of 3 acetone:1 Spurr's resin. Every 24 hours, the concentration of the resin was increased (2:1, 1:1, 0:1). The 100% Spurr's concentration was repeated three times for 24 hours each. The samples were cured overnight at 70°C. The polymerized blocks were hand trimmed and thin sectioned (80 nm) using a microtome and a glass knife. The thin sections, placed on a 200 mesh copper grid, were stained in uranyl acetate (4%) and Reynolds's lead citrate (0.17 M, pH 12). The samples were observed using TEM (JEOL 1200 EX II, Tokyo, Japan).

For SEM, the cells were sequentially dehydrated with 30, 50, 70, and 90% for 10 min, followed by two 15 min dehydration steps with 100% ethanol. Small pieces of the pellets and a couple of LWE cubes were placed on carbon tapes attached to metal stubs and gold coated for 3 minutes (Bozzola and Russell, 1992). The samples were observed using a Hitachi 570 scanning electron microscope (Hitachi, Tokyo, Japan).

COLOR MEASUREMENTS

Color was measured using a Minolta color spectrophotometer (Model CM2001, Minolta, Camera Ltd., Osaka, Japan). Whiteness was calculated according to Francis (1998): whiteness $= 100 - [(100 - L^*)^2 + a^{*2} + b^{*2}]^{1/2}$, where L^* denotes lightness on a 0 to 100 scale from black to white, a^* denotes $(+)$ redness or $(-)$ greenness, and b^* denotes $(+)$ yellowness or $(-)$ blueness.

RESULTS AND DISCUSSION

INACTIVATION RESULTS

PEF Treatment

The microbial population of stationary phase cells (Figure 13.4) of *P. fluorescens*, suspended in LWE, significantly decreased ($P < 0.05$) as the number of pulses increased. The inactivation of the three strains (ATCC 17400, ATCC13252, and WSU-07) by PEF treatment is presented in Figure 13.5. There is a significant difference ($P < 0.05$) between the inactivation rates of the three strains, showing that WSU-07 is more resistant to PEF treatment, with a total microbial reduction of 0.95 log cycles after 117 pulses. In contrast, ATCC 17400 was reduced by more than 3.5 logs. The third strain, ATCC 13525, was shown to be intermediate in resistance in comparison to the other two strains, with a final microbial reduction of more than 2 log cycles. Although there is a synergistic effect for the combination of PEF treatment with the addition to the LWE of 0.1% of parabens and 0.5% citric acid mixture (Figure 13.6), the effect of the antimicrobial mixture is not as important as the effect of the type of strain on the final microbial reduction (Figure 13.5). The hydrophobic characteristics of parabens that might cause some binding between them and the lipids of LWE can explain the limited synergistic effect of these antimicrobials. In order to overcome this, it might be necessary to increase the parabens mixture concentration or use a combination system of antimicrobials. The temperature profiles of all PEF treatments showed that the peak temperature during processing was

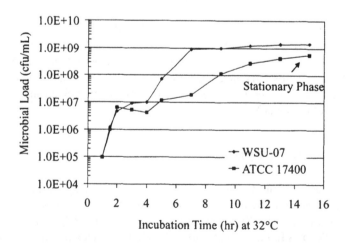

Figure 13.4 Growth curves of *P. fluorescens* strains in tryptic soy broth.

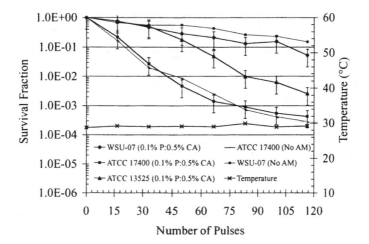

Figure 13.5 Inactivation of *P. fluorescens* strains by pulsed electric fields in liquid whole egg with parabens (P):citric acid (CA) mixture or no antimicrobial (No AM).

below 32°C (Figure 13.5), demonstrating that the inactivation of *P. fluorescens* in LWE by PEF treatment was due to a pulsing electric field effect.

The evaluation of the inactivation kinetics of the different strains gives a mathematical approach for comparison, and for predicting the needed treatment dosage in a food pasteurization process. An inactivation kinetic model, as proposed by Hülsheger et al. [1981: Equation (2)], can be used to correlate the survival fractions ($S = N/N_0$ = actual microbial count divided by the initial

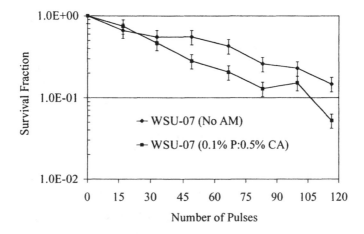

Figure 13.6 Inactivation of *P. fluorescens* WSU-07 by pulsed electric fields in liquid whole egg with parabens (P):citric acid (CA) mixture or no antimicrobial (No AM).

TABLE 13.2. Inactivation Kinetic Constants of *P. fluorescens* Inactivation by PEF.

			b		
Strain	ATCC 17400*	ATCC 17400**	ATCC 135252*	WSU-07*	WSU-07**
Parameter value	3.33	3.33	3.06	1.02	0.72
			$t_c(\mu s)$		
Parameter value	12.51	12.31	24.85	18.8	16.29
r^2	0.99	0.98	0.94	0.94	0.885

*Strain suspended in liquid whole egg with 0.1% parabens mixture and 0.5% citric acid.
**Strain suspended in liquid whole egg without antimicrobials.

microbial count) and the treatment time ($t = n^*\tau =$ number of pulses times the pulse width). Hülsheger's Equation (2) allows the evaluation of the survival rate constant (b) and the critical treatment time (t_c).

$$\ln(s) = -b \ln(t/t_c) \tag{2}$$

where b is the survival rate constant and t_c is the critical treatment time (an extrapolated value that corresponds to 100% survival). The kinetic parameters of *P. fluorescens* inactivation by PEF are listed in Table 13.2. The greater the value of the survival rate constant b, and the lower the critical time t_c, indicate a greater susceptibility of the *P. fluorescens* strain to the different PEF processing conditions.

The combination of PEF treatment and up to 1000 IU of nisin in the LWE (Figure 13.7) showed no significant ($P > 0.05$) synergistic effect on the

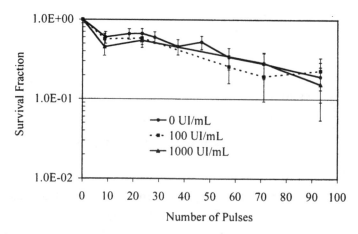

Figure 13.7 Inactivation of *P. fluorescens* WSU-07 by pulsed electric fields in liquid whole egg with nisin at different concentrations.

inactivation of *P. fluorescens*. These results suggest that PEF did not desta-
bilize the membrane of these gram-negative bacteria and nisin was not able to
penetrate. Furthermore, it is known that the effectiveness of antimicrobials is
lower in complex foods with high protein and fat contents, in comparison to
simpler systems such as microbiological media like TSB (Branen and Davidson,
1998).

HHP Treatment

To conduct the nonthermal HHP treatments, the most resistant WSU-07
and the least resistant ATCC 17400 *P. fluorescens* strains were selected. The
inactivation kinetics obtained by a 5 min HHP treatment at different pressures are presented in Figure 13.8. Pressure below 20 kpsi did not result in a significant inactivation, while pressure higher than 30 resulted in a
total inactivation (10^7 log cycles) of the two strains. In the range of 20 to
30 kpsi, both strains followed the same behavior, achieving an inactivation
of almost 4 log cycles. Thus, this pressure range was selected for further
study. It is worth mentioning that there are significant ($P < 0.05$) differences between the inactivation levels obtained for the two *P. fluorescens* strains. Furthermore, the ATCC 17400, the least resistant microorganism to the
PEF nonthermal treatment, was also the least resistant to HHP nonthermal
treatment.

The combination parabens (0.05–0.1%) and/or citric acid (0.15–0.5%) with
HHP treatment (5 min, 20 to 30 kpsi) (Figure 13.9) did not present a significant
($P < 0.05$) synergistic effect for either strain although a pairwise comparison

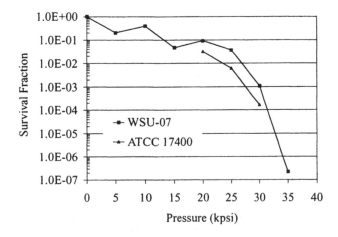

Figure 13.8 Inactivation of *P. fluorescens* by high hydrostatic pressure (5 min treatment).

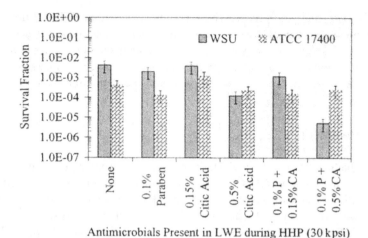

Figure 13.9 Effect of high hydrostatic pressure (30 kpsi, 5 min) combined with parabens and citric acid on the inactivation of *P. fluorescens* strains.

(Tukey's test) shows that citric acid, alone or in combination with parabens, yields a higher inactivation level.

The results for the inactivation of the WSU-07 strain by HHP (30 kpsi, 5 min) in combination with chemical and naturally occurring antimicrobials are presented in Figure 13.10. There are significant differences ($P < 0.0001$) among

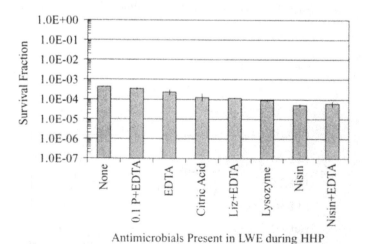

Figure 13.10 Effect of high hydrostatic pressure (30 kpsi, 5 min) combined with different antimicrobials on the inactivation of *P. fluorescens* WSU-07 strain.

the different antimicrobial combinations. Pairwise comparisons showed that nisin and lysozyme (with or without EDTA) are the most effective antimicrobials, yielding a significant synergistic effect with HHP treatment.

EVIDENCE OF ULTRASTRUCTURAL DAMAGE BY ELECTRON MICROSCOPY

Transmission electron microscopy micrographs of the WSU-07 and ATCC 17400 *P. fluorescens* strains, suspended in TSB before treatment, did not reveal internal structural differences (Figure 13.11). The high complexity of the LWE made it difficult to work with, although one electroplated cell of WSU-07 was found in a sample of LWE that had been PEF treated with 100 pulses of 48 kV/cm (Figure 13.12).

COLOR ANALYSIS

A PEF treatment of up to 117 pulses did not change the color of the LWE, as evidenced by no significant change in L^*, a^*, b^*, or whiteness (Figure 13.13). The increase in pressure used in the HHP treatments did not significantly change the redness of LWE, but decreased yellowness (data not shown), and

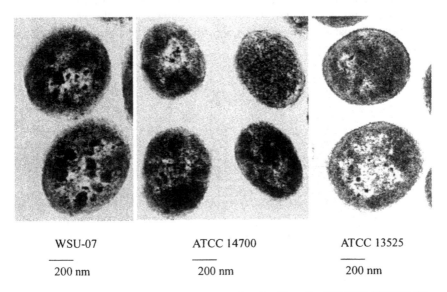

WSU-07	ATCC 14700	ATCC 13525
————	————	————
200 nm	200 nm	200 nm

Figure 13.11 Transmission electron microscopy graphs of healthy cells of WSU-07, ATCC 17400, and ATCC 13525 suspended in tryptic soy broth.

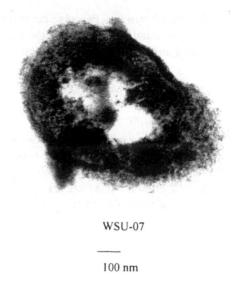

WSU-07

———

100 nm

Figure 13.12 Transmission electron microscopy graph of pulsed electric fields treated cells (100 pulses of 48 kV/cm) of WSU-07 suspended in liquid whole egg.

increased lightness (data not shown) and whiteness (Figure 13.14). The addition of citric acid significantly reduced the product's redness (Figure 13.13), but did not have a significant effect on its whiteness, lightness, or yellowness, although a pairwise comparison indicated a higher whiteness in those products with 0.5% citric acid (Figure 13.14).

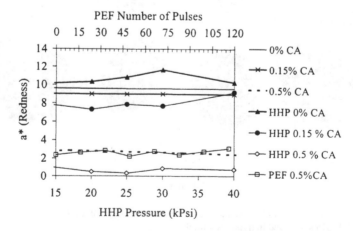

Figure 13.13 Effect of HHP or PEF and citric acid on the redness (a^*) of liquid whole egg.

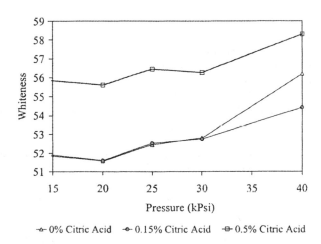

Figure 13.14 Effect of HHP and citric acid on the whiteness of LWE.

CONCLUSIONS

Nonthermal technologies demonstrated effectiveness in inactivating spoilage flora in food systems, yet preserving their high quality attributes. The hurdle approach had a significant synergistic effect on the inactivation of *P. fluorescens*. The limited results obtained with parabens can be explained by the hydrophobic characteristics of the antimicrobials, which might bind them to the lipids of LWE. In order to overcome this, it might be necessary to increase the antimicrobials concentration or to use a combination system of antimicrobials. High hydrostatic pressure had a significant impact on the gram-negative cell membranes, thus HHP sensitized the cell to the action of nisin and increased the effect of lysozyme. Also, it was observed that the WSU-07 strain was significantly more resistant to nonthermal treatments. In future commercial implementation of these processes, this *Pseudomonas* spp. may represent an indicator for the lethality evaluation of the process. The differences on the inactivation rates of the various strains suggest that care must be used to prevent selection of microbial flora during treatment.

Further studies are encouraged to explore the use of EDTA in higher concentrations (50 mM), as well as the use of monolaurin (500 to 1000 μg/mL) alone or in combination with lactic acid. The synergistic effect found with citric acid suggests that the use of other organic acids may obtain a higher inactivation rate.

REFERENCES

ATCC (American Type Culture Collection). 2000. *Online Catalog of Bacteriology*. American Type Culture Collection, Manassas, VA.

Barbosa-Cánovas, G. V., Góngora-Nieto, M. M. Pothakamury, U. R., and Swanson, B. G. 1999. *Preservation of Foods with Pulsed Electric Fields.* San Diego: Academic Press.

Barbosa-Cánovas, G. V., Pothakamury, U. R., Palou, E., and Swanson B. G. 1998. *Nonthermal Preservation of Foods. Food Science and Technology,* Vol. 82. New York: Marcel Dekker.

Bozzola, J. J. and Russell, L. D. 1992. *Electron Microscopy: Principles and Techniques for Biologists.* Boston, MA: Jones & Bartlett Publishers.

Branen, J. K. and Davidson, P. M. 1998. Enhanced activity of combinations of naturally occurring antimicrobials and membrane potentiators against food-borne pathogens. *IFT Annual Meeting,* Atlanta, GA, p. 107.

Code of Federal Regulations. 1997. Title 21. *Food and Drugs,* Parts 170–199. Washington, DC: Office of Federal Regulations, National Archives Record Service Administration.

Cutter C. N. and Siragusa, G. R. 1995. Population reductions of gram-negative pathogens following treatments with nisin and chelators under various conditions. *Journal of Food Protection.* 58 (9):977–983.

Davidson, P. M. 1992. Parabens and phenolic compounds. In: Davidson, P. M. and Branen, A. L. (editors), *Antimicrobials in Foods,* Second Edition. *Food Science and Technology,* Vol. 57, pp. 263–306. New York: Marcel Dekker.

Delves-Broughton, J., Williams, G. C., and Wilkinson, S. 1992. The use of the bacteriocin, nisin, as a preservative in pasteurized liquid whole egg. *Letters in Applied Microbiology* 15:133–136.

Doores, S. 1992. Organic acids. In: Davidson, P. M. and Branen, A. L. (editors), *Antimicrobials in Foods,* Second Edition. *Food Science and Technology,* Vol. 57, pp. 95–136. New York: Marcel Dekker.

Francis, F. J. 1998. Color analysis. In: Nelsen, S. S. (editor), *Food Analysis,* pp. 607–608. New York: Aspen Publishers.

Hauben, K. J. A., Wuytack, E. Y., Soontjens, C. C. F., and Michiels, C. W. 1996. High-pressure transient sensitization of *Escherichia coli* to lysozyme and nisin by disruption of outer-membrane permeability. *Journal of Food Protection.* 59 (4):350–355.

Ho, S. Y., Mittal, G. S., Cross, J. D., and Griffiths, M. W. 1995. Inactivation of *Pseudomonas fluorescens* by high voltage electric pulses. *Journal of Food Science.* 60 (6):1337.

Hülsheger, H., Potel, J., and Niemann, E. G. 1981. Killing of bacteria with electric pulses of high field strength. *Radiat. Environ. Biophys.* 20:53–65.

Liu, X., Yousef, A. E., and Chism, G. W. 1996. Inactivation of *Escherichia coli* O157: H7 by the combination of organic acids and pulsed electric fields. *Journal of Food Safety.* 16:287–299.

Martín-Belloso, O., Vega-Mercado, H., Qin, B. L., Chang, F. J., Barbosa-Cánovas, G. V., and Swanson, B. G. 1997. Inactivation of *Escherichia coli* suspended in liquid egg using pulsed electric fields. Journal of Food Processing and Preservation. 21:193–203.

National Agricultural Statistics Service. 1999a. Agricultural Statistics Board, U.S. Department of Agriculture. *Layers and Egg Production 1998 Summary.* http://usda.mannlib.cornell.edu/reports/nassr/poultry/pec-bbl/lyegan99.txt.

National Agricultural Statistics Service. 1999b. Agricultural Statistics Board, U.S. Department of Agriculture. *Egg Products.*

Palou, E. 1998. Food preservation by high hydrostatic pressure, process variables and microbial inactivation. Ph.D. Dissertation, Biological Systems Engineering Department, Washington State University, Pullman, WA.

Qin, B. L., Pothakamury, U. R., Barbosa-Cánovas, G. V., and Swanson B.G. 1996. Non thermal pasteurization of liquid foods using high intensity pulsed electric fields. *Critical Reviews in Food Science and Nutrition.* 36 (6):603–627.

Qin, B. L., Barbosa-Cánovas, G. V., Swanson, B. G., Pedrow, P. D., and Olsen R. G. 1998. Inactivating microorganisms using a pulsed electric field continuous treatment system. *IEEE. Transactions on Industry Applications*. 34 (1):43–49.

Qin, B. L., Zhang, Q., Barbosa-Cánovas, G. V., Swanson, B. G., and Pedrow, P. D. 1995. Pulsed electric field treatment chamber design for liquid food pasteurization using a finite element method. *Transactions of the ASAE*. 38 (2):557–565.

Vanderzant, C. and Splittstoesser, D. F. 1992. *Compendium of Methods for the Microbiological Examination of Foods*, Third Edition. Washington, DC: American Public Health Association.

CHAPTER 14

Reformulation of a Cheese Sauce and Salsa to Be Processed Using Pulsed Electric Fields

K. T. RUHLMAN
Z. T. JIN
Q. H. ZHANG
G. W. CHISM
W. J. HARPER

ABSTRACT

CHEESE sauce and salsa are not suitable for PEF treatment due to their salt content, viscosity, and particle size. Salsa and cheese sauce products were reformulated so that the electrical conductivity was minimized, and the particle size was reduced. For each product, the electrical conductivity, density, and viscosity were evaluated at temperatures ranging from 4°C to 80°C. Energy input and temperature rise during PEF processing, so pumping requirements were calculated for temperatures from 4°C to 80°C. The reformulated products were suitable for PEF processing.

INTRODUCTION

Pulsed electric field (PEF) processing is a nonthermal method used to increase shelf life and assure food safety by inactivating spoilage and pathogenic microorganisms. This method of preservation minimizes the increase in product temperature during processing, which maintains the flavor, color, and nutritional value of the product (Dunn and Pearlman, 1987; Jin and Zhang, 1999; Jin et al., 1998; Jia et al., 1999). This process is used in combination with, or in replacement of, thermal sterilization methods. To process a food with PEF in a continuous system, the food flows through a series of treatment zones, with a high voltage electrode on one side of each zone, and a low voltage electrode on the other. The PEF process is defined by the electric field strength, or the voltage applied per distance between electrodes, and the treatment time.

213

This method has proven successful in a variety of liquid products with a relatively low viscosity and electrical conductivity (Zhang et al., 1994; Jin et al., 1998; Mizuno and Hori, 1988). However, PEF has not yet been successfully used for processing products with high electrical conductivity, high viscosity, or large particle size.

The physical properties of the foods being processed, including electrical conductivity, viscosity, and density, are very important when designing and operating a PEF process (Ruhlman et al., 2001). Operational conditions need to be adjusted to accommodate products with a high electrical conductivity and viscosity, like cheese sauce and salsa. These adjustments include: PEF process parameters, fluid handling system design, and the physical properties of the food. The presence of salt contributes to the high electrical conductivity of cheese sauce and salsa. By preparing the products to be PEF processed without salt, the electrical conductivity can be reduced to meet the limitation of our PEF system. The appropriate concentration of salt can be aseptically added to the product after PEF processing.

The objective of this study was to reformulate salsa and cheese sauce to meet the requirements of the PEF system in terms of electrical conductivity and viscosity. This is the first step toward successfully using PEF for processing highly viscous foods or foods with large particles.

METHODS AND MATERIALS

For this study, the salt indicated in Table 14.1 (salsa) and Table 14.2 (cheese sauce) was omitted from the formulation until after PEF treatment to eliminate excess conductivity.

SALSA

Two salsa products were formulated using fresh ingredients, spices, and salt (Table 14.1). To process particulate foods, the treatment chamber was

TABLE 14.1. Two Salsa Formulations.

Salsa #1	Salsa #2
88% Tomatoes	69% Tomatoes
5.5% Cilantro	12% Green peppers
2.8% Lemon juice	12% Green onions
1.7% Green onions	3.6% Lemon juice
0.9% Cayenne pepper sauce	2.9% Jalapeño chilees
0.5% Cumin	0.5% Crushed garlic
0.5% Crushed garlic	[0.7% Salt]
[1% Salt]	

[ ] Indicates that salt is to be added after PEF processing.

TABLE 14.2. Two Cheese Sauce Formulations.

Cheese Sauce #1	Cheese Sauce #2
81.25% De-ionized water	81.25% De-ionized water
15% Protein (18.75% MPC)	15% Protein (18.75% MPC)
0.75% Sodium citrate	0.50% Sodium citrate
0.5% Cheddease 715	0.5% Cheddease 715
[1% Salt]	[1% Salt]

[ ] Indicates that salt is to be added after PEF processing.

redesigned to provide an inner diameter of 0.635 cm. The size of the particles in the salsa was reduced to approximately 1/8 in. (0.317 cm) by blending in a Waring blender. Since salt influences the electrical conductivity of the food, the formulations were prepared without added salt. The electrical conductivity and apparent viscosity were measured at 4°C, 22°C, 30°C, 40°C, 50°C, 60°C, 70°C, and 80°C.

CHEESE SAUCE

Two cheese sauce products were formulated using milk protein concentrate (Alapro 4850, New Zealand Milk Products, Inc.), de-ionized water, sodium citrate, salt, and enzyme modified cheese (Land O'Lakes Cheddease 715 Cheese Powder) (Table 14.2). After the sodium citrate was added, the mixture was heated to 60°C to aid emulsification, and then cooled to room temperature. The two formulations differ only in the amount of sodium citrate added, because sodium citrate affects both the viscosity and the electrical conductivity of the product. The electrical conductivity and apparent viscosity were measured at 4°C, 22°C, 30°C, 40°C, 50°C, and 60°C. In order to see the effects of the sodium citrate and added heat on the apparent viscosity, three sets of data were gathered: all ingredients, except sodium citrate, were mixed and the product was refrigerated before viscosity measurement; all ingredients, including sodium citrate, were mixed and the product was refrigerated before viscosity measurement; and all ingredients were mixed, heated to 60°C, and then refrigerated before viscosity measurement.

MEASUREMENTS OF PHYSICAL PROPERTIES

The electrical conductivity of both products was measured using a Yellow Springs conductivity meter (model 30/10 FT, YSI Incorporated, Yellow Springs, OH). The apparent viscosity of the salsa was measured using a Brookfield rotational viscometer (model DV-II+ with spindle LV #1 at 30 and 100 RPM, Brookfield Engineering Laboratories, Stoughton, MA). The apparent viscosity of the cheese sauce was measured using a Brookfield HBDVII+ cone and plate viscometer with spindle cp-40 at shear rates of 75, 350, and 375 s^{-1} (Brookfield

Engineering Laboratories, Stoughton, MA). For both products, the density was measured at room temperature using a Fisher Scientific pycnometer (Fisher Scientific, Pittsburgh, PA), and the specific heat was calculated using a model estimation for materials of high water content (Singh and Heldman, 1993).

CALCULATIONS

The maximum possible temperature change per pair of treatment chambers (ΔT), and total energy input during treatment per pair of chambers (P), were calculated for both products at all temperatures using the following equations:

$$\Delta T = (E^2 t\sigma / \rho C_p)/n$$
$$P = E^2 t\sigma / n$$

The variables in these equations are electrical conductivity (σ), density (ρ), and specific heat (C_p), and they are specific for each product. For the calculations, the PEF system designed for processing orange juice was used as an example (Zhang, 1997), where:

E = electric field strength (3.2×10^6 V/m)
t = total treatment time (9.0×10^7 s)
n = number of pairs of treatment chambers (6)
D = treatment zone diameter (4.8×10^{-3} cm)

The viscosity was used to determine the pumping requirements in the treatment chamber, which were calculated using a microsoft Excel spreadsheet designed specifically for our pilot plant design.

RESULTS AND DISCUSSION

SALSA

Commercial salsa products that contain salt, have an electrical conductivity of about 2.5 S/m, approximately 7 times higher than that of orange juice (0.35 S/m). Products with higher electrical conductivity require more energy input and better temperature control during PEF processing. The current PEF process system cannot meet the requirements. Since salt is a major contributor to the electrical conductivity of the product, the best solution would be to eliminate the salt from the product. However, with salsa, the acceptability of the product flavor is compromised. We suggest that salsa should be prepared without salt for PEF processing. Before packaging, the salt can be aseptically added to the salsa as a sterile salt solution.

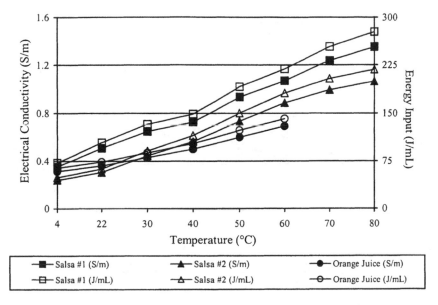

Figure 14.1 Electrical conductivity and calculated energy input per pair of chambers vs. temperature for salsa.

With an increase in temperature, there is an increase in the electrical conductivity, and the energy input is directly proportional to the electrical conductivity of the food (Figure 14.1). Since the only variable in the calculation of energy input is the electrical conductivity, this was expected. The electrical conductivity values for the salsas before salt was added are both about two times greater than for the orange juice. Currently, orange juice is PEF processed at 30°C. At this temperature, the energy input is approximately 87 J/mL per pair of chambers. When the salsa products are processed at 40°C, the energy consumption per pair of chambers would be 148 J/mL for salsa #1 and 115 J/mL for salsa #2.

The change in temperature during treatment is directly proportional to the electrical conductivity of the product (Figure 14.2). If salsa #2 enters the first pair of chambers at 40°C, there will be a maximum of 27°C increase in the product temperature when it exits. This temperature increase needs to be carefully controlled because as the temperature increases, the electrical conductivity and energy consumption also increase. By using a cooling heat exchanger between each pair of chambers, this temperature change is kept to a minimum.

The apparent viscosity of both salsa reformulations decreases with an increase in temperature (Figure 14.3). These products are pseudoplastic, which means that the viscosity decreases with an increase in shear rate. The viscosity was measured at very low shear rates compared to what the product will

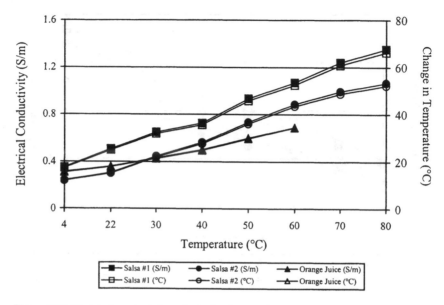

Figure 14.2 Electrical conductivity and calculated energy input per pair of chambers vs. temperature for salsa.

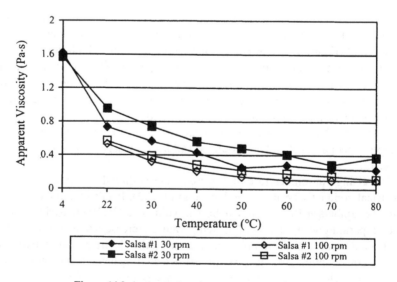

Figure 14.3 Apparent viscosity vs. temperature for salsa.

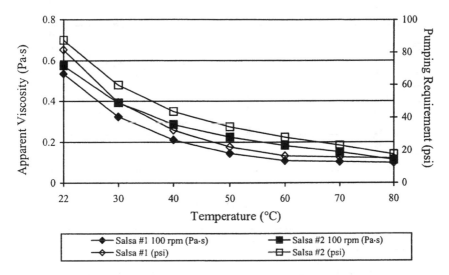

Figure 14.4 Apparent viscosity and calculated pumping requirement for the chamber vs. temperature for salsa.

encounter during processing. If the product is flowing at a very high rate of shear during treatment, the viscosity will be reduced even further. As the temperature increases, the viscosity and pumping requirements decrease (Figure 14.4). This is important because of the limitation of the fluid handling system. We suggest that the pressure not exceed 100 psi for the entire fluid handling system if a pump with maximal capacity of 150 psi is used. Since the treatment chamber has the smallest inner diameter of all of the tubing of the fluid handling system, the greatest pressure drop will be seen through this section.

CHEESE SAUCE

Removing all of the salt from the cheese sauce, including the sodium citrate and salt added for flavor, will reduce the electrical conductivity. However, this creates a new problem with the product viscosity, since sodium citrate is added to aid emulsification, and to enhance the texture of the product. With an increase in the amount of sodium citrate, there is an increase in the electrical conductivity of the cheese sauce (Figure 14.5). The electrical conductivity of the cheese sauce with 0.5% sodium citrate was less than the electrical conductivity of the orange juice. Since it is known that the energy input is directly proportional to the electrical conductivity of the food, this indicates that less energy will be required to process the cheese sauce than to process the orange juice.

As the input temperature is increased, the temperature change is increased (Figure 14.6). The change in temperature during treatment is proportional to

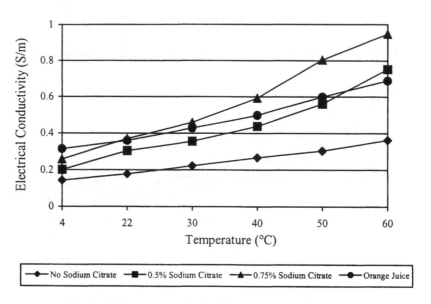

Figure 14.5 Electrical conductivity vs. temperature for cheese sauce.

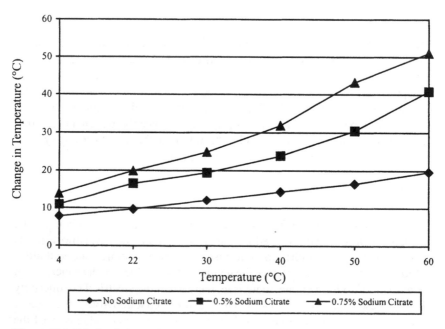

Figure 14.6 Calculated change in temperature per pair of chambers vs. input temperature for cheese sauce.

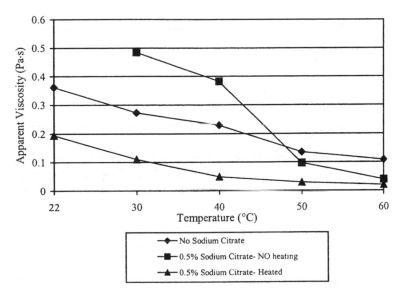

Figure 14.7 Apparent viscosity at 375 s^{-1} shear rate for cheese sauce vs. temperature.

the electrical conductivity of the product, and can be kept to a minimum with the use of a cooling heat exchanger.

The 0.5% sodium citrate cheese sauce that was heated to 60°C and cooled before measuring had the lowest apparent viscosity (Figure 14.7). Also, at 50°C the viscosity dropped rapidly for the products containing sodium citrate, and then changed very little at 60°C. This indicates that 50°C would be the optimum treatment temperature in terms of viscosity. When measured at three rates of shear, the cheese sauce exhibited pseudoplastic properties (Figure 14.8). This indicates that as the product moves faster through the tubing, it will encounter more shear rate, and the viscosity will be further reduced. As the viscosity is decreased, the pressure differential is also decreased (Figure 14.9). At 50°C the pressure differential is approximately 5 psi in the treatment chamber, which has the smallest inner diameter of the entire fluid handling system.

CONCLUSIONS AND RECOMMENDATIONS

In order to have successful PEF treatment of products with high electrical conductivity and viscosity, a combination of process parameter adjustment and product physical property adjustment is necessary. Also, careful control of temperature during treatment is necessary to assure that the desired dosage levels are achieved.

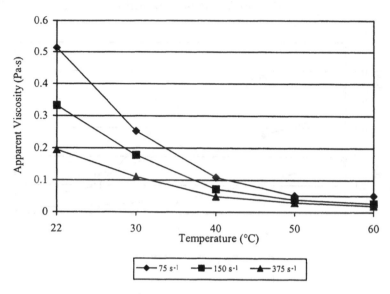

Figure 14.8 Apparent viscosity at increasing shear rate vs. temperature for cheese sauce containing 0.5% sodium citrate.

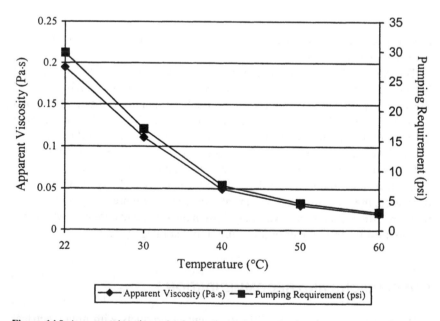

Figure 14.9 Apparent viscosity and calculated pumping requirement vs. temperature for cheese sauce with 0.5% sodium citrate.

222

We recommend that for PEF processing, salsa #2 be prepared without salt and blended until the particles are less than 1/8 inch in size, if a chamber size of 0.635 cm ID (½ in.) is used. In order to be most efficient in terms of pumping requirements and energy input, it is recommended that processing begin at 40°C. At 40°C the energy input needed to achieve a 32 kV/cm electric field strength is 115 J/mL, and the temperature change is 30°C per pair of treatment chambers. Also, the pumping requirement is 30 psi, which is approximately 30% of the total pumping limits.

We recommend that a cheese sauce containing 0.5% sodium citrate be prepared and heated to 60°C before PEF processing. Processing at 50°C is recommended for efficiency. The energy input needed to achieve a 32 kV/cm electric field strength is 114 J/mL, and the temperature change is 30°C per pair of chambers. The pumping requirement for the treatment chamber is 5 psi, which is 5% of the total pumping limitations.

REFERENCES

Dunn, J. E. and Pearlman, J. S. 1987. Methods and apparatus of extending the shelf life of fluid food products. U.S. patent 4,695,472.

Jia, M., Zhang, Q. H., and Min, D. B. 1999. Pulsed electric field processing effects on flavor compounds and microorganisms of orange juice. *Food Chemistry*, 65(4):445.

Jin, Z. T., Ruhlman, K. T., Qiu, X., Jia, M., Zhang, S., and Zhang, Q. H. 1998. Shelf life evaluation of pulsed electric fields treated aseptically packaged cranberry juice. *98 IFT Annual Meeting*. Atlanta, GA. June 20–24. Book of Abstracts p. 70.

Jin, Z. T. and Zhang, Q. H. 1999. Pulsed electric field treatment inactivates microorganisms and preserves the quality of cranberry juice. *J. Food Proc. Pres.*, 23(6):481–49

Mizuno, A. and Hori, Y. 1988. Destruction of living cells by pulsed high voltage application. *IEEE Trans. on Indus. Appli.*, 24(3):387–394.

Ruhlman, K. T., Jin, Z. T., and Zhang, Q. H. 2001. Physical properties of liquid foods for pulsed electric field treatment. Chapter 3 in: *Pulsed Electric Fields in Food Processing: Fundamental Aspects and Applications*. G. V. Barbosa-Cánovas and Q. H. Zhang, Eds. Technomic Publishing Co., Inc., Lancaster, PA.

Singh, R. P. and Heldman, D. R. 1993. *Introduction to Food Engineering*. Academic Press Inc., San Diego.

Zhang, Q. H. 1997. Continuous PEF systems. *The Second PEF Workshop*. Chicago, IL. October 1997.

Zhang, Q. H., Monsalve-González, A., Qin, B. L., Barbosa-Cánovas, G. V., and Swanson, B. G. 1994. Inactivation of *Saccharomyces cerevisiae* in apple juice by square wave and exponential-decay pulsed electric fields. *J. Food Proc. Eng.*, 17:469–478.

CHAPTER 15

Comparison Study of Pulsed Electric Fields, High Hydrostatic Pressure, and Thermal Processing on the Electrophoretic Patterns of Liquid Whole Eggs

L. MA
F. J. CHANG
M. M. GÓNGORA-NIETO
G. V. BARBOSA-CÁNOVAS
B. G. SWANSON

ABSTRACT

THE effects of thermal and nonthermal [pulsed electric fields (PEF) and high hydrostatic pressure (HHP)] processing on proteins of liquid whole eggs (LWE) were studied by native polyacrylamide gel electrophoresis (native PAGE). The PEF process, with up to 50 pulses, had little effect on the protein bands identified by native PAGE. However, electrophoretic protein pattern changes were observed when the pressure was greater than 50 kpsi. The protein bands, such as γ-livetins and ovomacroglobulins, were the most pressure sensitive. Some protein bands started fading when the temperature was greater than 60°C. More faded conalbumins, globulins, ovomacroglobulin, and γ-livetins were observed at 70°C.

INTRODUCTION

Eggs are among the most complete foods available to humans that are relatively inexpensive (Ronsivalli and Vieira, 1992). Although the contents of freshly laid eggs are generally sterile, shell surfaces contain many bacteria. Liquid or raw beaten eggs may be contaminated during processing with microorganisms, such as *E. coli* because of improper handling and unsanitary conditions

(Vanderzant and Splittstoesser, 1992). Thus, liquid eggs in the United States are currently pasteurized (Foegeding and Stanley, 1987) or ultrapasteurized (Ball et al., 1987), and stored refrigerated or frozen to inactivate *Salmonella* and prolong shelf life. Regulatory agencies in the United States and several other countries require pasteurization of commercial egg products that are removed from the shells. Pasteurization of whole eggs requires heating at 60°C to 62°C for 3.5 to 4 min (Potter and Hotchkiss, 1995). Successful pasteurization is based on a critical time-temperature relationship because a lower than specified temperature decreases the efficiency of pasteurization, and overheating may result in coagulation of the egg and formation of a film on the heat exchanger surface (Banwart, 1989; Powrie and Nakai, 1985).

Changes in both the physical and functional properties of egg white due to heat have been described by many early researchers (Ball et al., 1987; Brant et al., 1968; Cotterill et al., 1976; Dixon and Cotterill, 1981; Hamid-Samimi and Swartzel, 1985; Matsuda et al., 1981; Woodward and Cotterill, 1983). It has been reported that heating egg whites in the pasteurization range of 54°C to 60°C (129°F to 140°F) damages their foaming power (Cunningham, 1995). Bakery workers agreed that egg whites are impaired when heated for several minutes above 57°C. Egg whites (at pH 9) tend to increase in viscosity when heated to 56.7°C to 57.2°C, and coagulate rapidly at 60°C. Whole eggs pasteurized in the temperature range of 60°C to 68°C produce sponge cakes with volumes approximately 4% less than those made from control samples. Cakes made with whole eggs pasteurized at 71°C have volumes 8% lower than those made from control samples (Cunningham, 1995). Poor texture has also been noted in sponge cake made from pasteurized whole eggs. Such limitations in thermal treatments for the pasteurization of LWE make it necessary to consider a nonthermal procedure to inactivate microorganisms in egg products.

The inactivation of microorganisms in food systems by PEF and HHP was first studied by Sale and Hamilton (1967) and Hite (1989), respectively. Matsumoto et al. (1990) reported that no change in the taste or fragrance of sake resulted from PEF processing. Sensory evaluations of milk and orange juice by Grahl et al. (1992) indicated that taste did not deteriorate significantly after PEF treatment. Food processing with HHP is on the verge of becoming accepted worldwide. Japanese researchers have published several articles reporting that HHP treated food retains its fresh taste and flavor. By producing the first food products processed by HHP available to consumers in April of 1990 (Rovere, 1995), the Japanese are considered the current world leaders in this type of food processing. Increasing shelf life by a few days with HHP may be considered successful, because foods that typically spoil within days could be distributed in much wider ranges. An example is the processing of mackerel by HHP, which results in a four day extension of shelf life, as judged by freshness indices and sensory evaluations (Fujii et al., 1994).

Production of LWE products with extended nonfrozen shelf lives would reduce spoilage losses, allow for a greater range of distribution, and improve the safety of the product. PEF and HHP are nonthermal processing techniques being developed to extend the nonfrozen shelf-life of LWE. Despite the advantages of nonthermal processing, its effect on egg protein functionality is of concern to processors. Many researchers have evaluated the effects of heat treatments greater than the U.S. minimum pasteurization requirement (60°C for 3.5 min) using PAGE as a means to detect heat induced protein changes (Chang et al., 1970; Torten and Eisenberg, 1982; Woodward and Cotterill, 1983; Watanabe et al., 1986). In these studies the batch heating of LWE in test tubes was used as the means of treatment. Egg samples were usually held at a constant heating time of 3 to 3.5 min. Comparatively little has been published on the effects of nonthermal processing techniques on the proteins of LWE. The objectives of our study were to compare the proteins of LWE when processed by thermal and nonthermal (HHP and PEF) techniques using native PAGE. Another potential use of native PAGE was to identify specific bands associated with egg functionality or adequate heat treatment.

MATERIALS AND METHODS

LWE

Fresh eggs from a local supermarket were inspected for shell integrity, rinsed with distilled water, soaked in 70% alcohol for 3 min before breaking their shells, removed of their contents, and stored in a sterile beaker until 4.0 L was collected. The LWE were beaten using a Hobart mixer (Hobart Manufacturing Co., Troy, OH) for 10 min at the lowest speed. Ten milliliters of 60% (w/v) citric acid was added between the second and third min. The homogeneous LWE was filtrated twice with a kitchen screen sifter. The above procedure was repeated for each of the experiments.

PEF

A continuous treatment chamber (Figure 15.1), consisting of a concentric electrode and stainless steel body with 28 mL capacity and 0.6 cm gap, was used to apply the high intensity PEF treatments with a constant flow rate of 500 mL/min, using a peristaltic pump (Masterflex Model 7564-00, Cole Parmer Instrument Co., Chicago, IL). The pulsing rate was set at 3 Hz with a 40 kV input voltage (Table 15.1).

The electric field intensity was determined by an oscilloscope (Hewlett Packard 54520A, Colorado Springs, CO), and the electric field generated using a pilot plant size pulser (Physics International, San Leandro, CA). The

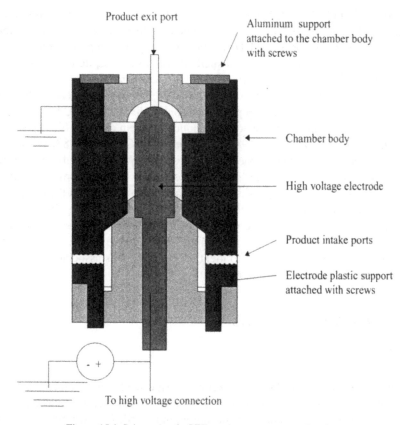

Product exit port

Aluminum support
attached to the chamber body
with screws

Chamber body

High voltage electrode

Product intake ports

Electrode plastic support
attached with screws

To high voltage connection

Figure 15.1 Schematic of a PEF continuous treatment chamber.

temperature of the treated egg was observed at the exit of the treatment chamber
using a digital thermometer (John Fluke Mfg. Co., Everett, WA).

Gas bubbles were removed from the treatment chamber after the addition
of LWE, and prior to the application of PEF to reduce the possibility of arcing. LWE was then pumped through the chamber while pulsing in a stepwise
treatment mode. Four pulses were applied in each step, with five consecutive
steps used to treat the liquid egg (Figure 15.2). The treated product was collected and cooled to 15°C after each 10 pulse treatment, when the treatment
chamber and fittings were dissembled and cleaned thoroughly using chlorine
(>200 ppm) and sterilized water. The liquid egg from the previous step was
retreated.

The PEF operation was run at room temperature (−22°C), with an electric
field of 48 kV/cm, capacitance of 0.5 μF, and exponential decay pulses of ~2 μs

TABLE 15.1. Treatment Conditions for LWE Exposed to PEF.

	Operation Conditions
Capacitance (VF)	0.5
Input voltage (kV)	40
Input flow rate (L/min)	0.5
Input pulse rate (Hz)	3
Peak voltage (kV)	29
Electric field intensity (kV/cm)	48
Pulse energy (J)	210
Maximum temperature (°C)	40

utilized. The PEF processed LWE received 10 pulses per unit volume. After treatment the product was poured into 1 L Scholle bags (type I L. MET. W/800R IRR, Scholle Co., Northlake, IL).

HHP TREATMENTS

One liter of LWE was poured into each 1 L Scholle bag. The Scholle bags were placed in polyethylene pouches (Power Plastics Inc., Paterson, NJ), and

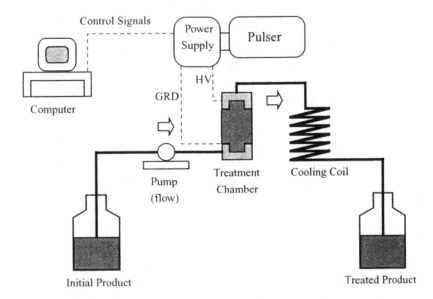

Figure 15.2 A single pass PEF operation.

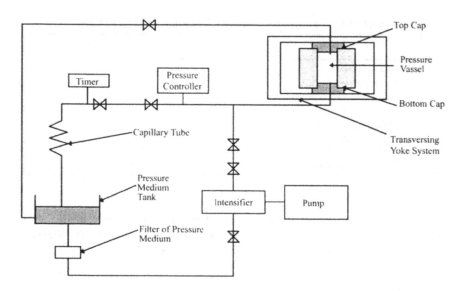

Figure 15.3 Schematic diagram of an HHP system.

100 mL of sterile distilled water was added to each pouch. This was followed by removal of as much air as possible, and then heat scaling. A pilot plant scale hydrostatic pressure system (Figure 15.3) (Engineered Pressure Systems Inc., Andover, MA), operated at room temperature ($-22°C$), was used to process the Scholle bags and polyethylene pouches of LWE. The applied pressures were set at 0, 30, 40, 50, 60, 70, or 80 kpsi with a holding time of 5 min. The holding times were set for 0, 1, 3, 5, 7, or 9 min at 60 kpsi.

THERMAL PROCESSING

Water baths set at 55, 60, 65, or 70°C were used to heat the LWE, which was then put into a semicapillary heating tube ($\phi = 5$ mm, $L = 85$ mm), and placed into the water baths at 55, 60, 65, or 70°C for 3 min. At the end of the holding time (3 min), the capillary tube was immediately transferred into an ice bath to cool.

SAMPLES

One-half milliliter of thoroughly mixed LWE was added to 5.5 mL of a diluted buffer mixture composed of a 50 mM solution of neutral Tris buffer, containing 8% glycine and 0.025% Bromophenol Blue. The egg buffer mixture was combined in a glass test tube. Nine microliters of this mixture, with 90 micrograms protein, were applied to the gel.

ELECTROPHORESIS

Native PAGE was performed on all control and heated egg samples from three replications, using a Hoefer vertical slap apparatus (model SE600, Hoefer Scientific Instrument, San Francisco, CA) and Phamacia power supply (model ECPS 300/150, Phamacia Electrophoresis Constant Power Supply). A 10% running gel and 3.5% stacking gel were made following the procedure of Laemmli (1970), excluding SDS. The samples were run at a constant current of 40 mA. The gel was stained in 0.12% Coomassie Blue for ~24 hr and destained in a 7% glacial acetic acid and 40% methanol solution for ~18 hr. The protein bands were identified by comparing them to native gels found in the literature (Torten and Eisenberg, 1982).

RESULTS AND DISCUSSION

PHYSICAL APPEARANCE

The appearance of the LWE processed by HHP technology began to change at 50 kpsi, when it became slightly white. Soft curds formed at higher pressures (Ibarz et al., 1996). The appearance of heated LWE became slightly opaque at 60°C. A soft coagulum formed at 75°C, which was in agreement with data from the literature (Woodward and Cotterill, 1983). However, the appearance of LWE processed by PEF did not change with up to 50 pulses at 48 kV/cm intensity.

EFFECT OF HEATING ON THE ELECTROPHORETIC PATTERNS OF LWE PROTEINS

Native PAGE separation of LWE without any treatment gave 16 distinguishable protein bands (Lane 2 in Figure 15.4). Standard protein served as a reference (Lane 1 in Figure 15.4). Bands were identified according to previous studies (Chang et al., 1970; Dixon and Cotterill, 1981; Matsuda et al., 1981; Woodward and Cotterill, 1983; Torten and Eisenberg, 1982). Figure 15.5 shows a tentative electrophoretic protein band diagram according to the literature (Torten and Eisenberg, 1982). The egg protein is stable at 55°C, since there was no band fading (Lane 3 in Figure 15.4). Compared to the control LWE, the conalbumin band started fading at 60°C (Lane 4 in Figure 15.4), which is in agreement with the results reported by many researchers (60°C to 65°C) (Chang et al., 1970; Donovan et al., 1975; Hegg et al., 1978; Nakamura et al., 1979). When the temperature increased to 65°C, more faded conalbumins were observed, and some globulins, ovomacroglobulin, and γ-livetin started fading (Lane 5 in Figure 15.4). Ovomacroglobulin (OMG) is a large egg white protein that

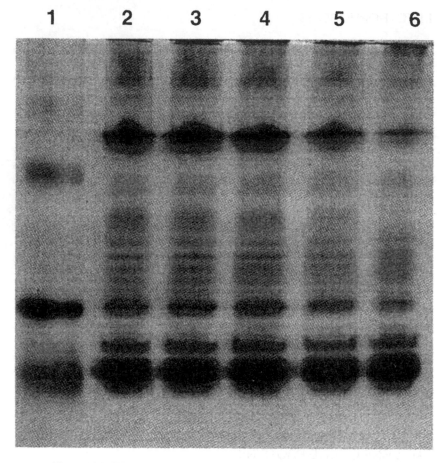

Figure 15.4 Effect of heating on the electrophoretic patterns of LWE proteins.

barely migrates into gels due to its size (6.5×10^5 MW), and its stability can be increased to 63°C and 78°C by the addition of sugar or salt (NaCl) (Woodward and Cotterill, 1983). More faded conalbumins, globulins, ovomacroglobulin and γ-livetin were observed at 70°C (Lane 6 in Figure 15.4).

EFFECT OF HHP ON THE ELECTROPHORETIC PATTERN OF LWE PROTEINS

The effects of HHP on the electrophoretic pattern of LWE proteins are presented in Figure 15.6. Lane 2 in Figure 15.6 represents the electrophoretic pattern of LWE without any treatment. This served as the control for the comparison. Standard protein was used as the reference (Lane 1 in Figure 15.6).

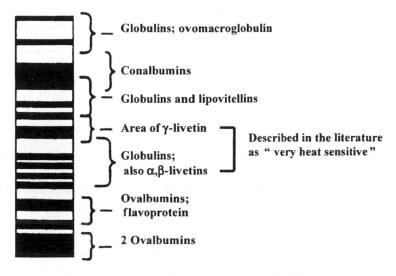

Figure 15.5 Electrophoretic patterns of non-treated LWE proteins.

There were no protein band changes at pressure 30 kpsi (Lane 3 in Figure 15.6). When the pressure increased to 40 kpsi, some OMG started fading (Lane 4 in Figure 15.6). As the pressure increased, more and more OMG faded, and almost all had done so at 70 kpsi (Lane 8 in Figure 15.6). OMG was thus determined as not only heat sensitive (Woodward and Cotterill, 1983), but also pressure sensitive. At a pressure of 60 kpsi, some γ-livetin started fading (Lane 6 in Figure 15.6). Almost all γ-livetin and β-livetin faded at 60 or 70 kpsi (Lanes 7 and 8 in Figure 15.6).

HHP is considered a nonthermal processing technology, since it requires a temperature rise of only 3°C per 15 kpsi when treating products high in water content (Rovere, 1995). Therefore, the heat generated during HHP was not responsible for the change in LWE protein bands, since the maximum temperature rise in our study was less than 15°C. It is not clear how high pressure altered the egg protein. However, it is possible that the volume change due to the high pressure compression was a key factor in the alteration of some LWE protein bands (Table 15.2), since the volume was compressed up to 22% at 80 kpsi pressure. It is known that proteins exist in solutions with different conformation (second and third order structures) with the help of hydrogen bonds, van der Waals force, hydrophobic interaction, and interaction with surrounding water molecules. When the volume is reduced, the freedom of these interactions is significantly restrained. Therefore, the higher order structure (conformation) of some proteins in LWE may be altered due to the compression of volume by high pressure. From this study, the critical volume compression was in the range of 10% to 15%, which corresponds to pressures of 50 to 60 kpsi. When the

TABLE 15.2. Volume Compression in HHP.

Applied Pressure (kpsi)	Volume Decreased (%)
0	0
30	8.9
40	11.7
50	14.4
60	17.0
70	19.5
80	22.0

pressure was greater than the critical pressure, the three-dimensional structure and conformation changed.

Figure 15.7 illustrates the effect of holding time on the electrophoretic pattern of LWE at a pressure of 60 kpsi. Lane 2 in Figure 15.7 represents the electrophoretic pattern of LWE without any treatment. This served as the control

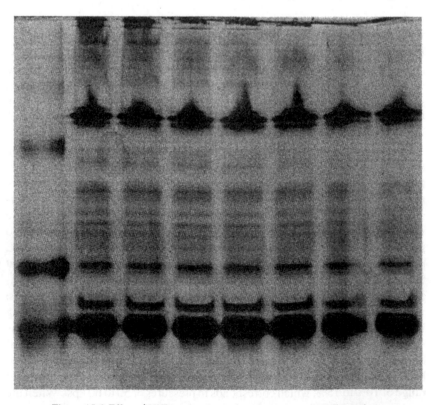

Figure 15.6 Effect of HHP on the electrophoretic patterns of LWE proteins.

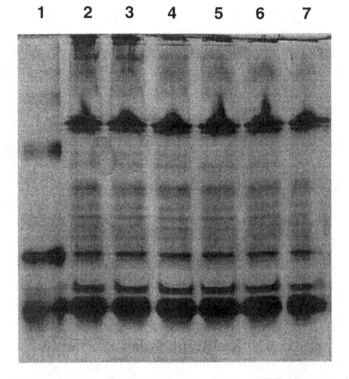

Figure 15.7 Effect of holding time on the electrophoretic patterns of LWE at a pressure of 60 kpsi.

for the comparison. As such, only OMG faded. The electrophoretic patterns under different holding times were found to be very similar (Lane 3 to 7 in Figure 15.7). Therefore, it may be concluded that the holding time is not critical in changing protein patterns, but pressure is.

EFFECT OF PEF ON THE ELECTROPHORETIC PATTERN OF LWE PROTEINS

The effects of PEF on the electrophoretic pattern of LWE proteins are presented in Figure 15.8. Lane 2 in Figure 15.8 represents the electrophoretic pattern of LWE without any treatment. This served as the control for the comparison. Standard protein was used as the reference (Lane 1 in Figure 15.8). In comparing the electrophoretic protein bands of all LWE processed by PEF (Lanes 3 to 7 in Figure 15.8) to the control LWE (Lane 2 in Figure 15.8), it is noted that there were no band changes up to 50 pulses. With PEF, the processing temperature was maintained below 40°C. Therefore, the thermal effects due to PEF processing were minimized. In addition, the total PEF processing time was

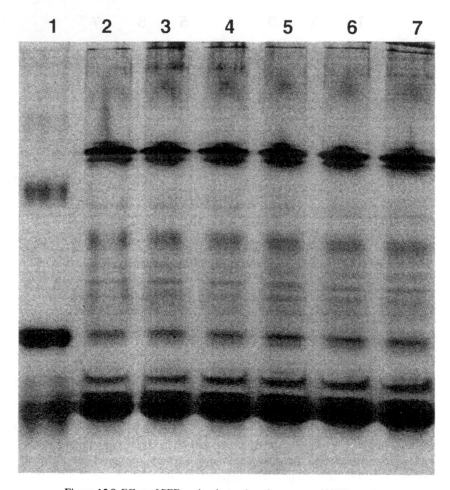

Figure 15.8 Effect of PEF on the electrophoretic patterns of LWE proteins.

very short (microseconds), and can be computed by the following equation:

$$t = n\tau = nRC \tag{1}$$

where

t = total processing time (s)
n = pulse number
τ = duration of a single pulse
R = effective resistance of LWE in the processing chamber
C = the capacitance of pulser

TABLE 15.3. The Processing Time in PEF Treatment.

Applied Number of Pulses	Processing Time (Microsecond)
0	0
10	23.9
20	47.8
30	71.7
40	95.6
50	119.5

In our study, the effective resistance of LWE and the pulser capacitance were 4.78 Ω and 0.5 μF, respectively. Therefore, processing under different pulses can be calculated using Equation (1) (see Table 15.3). The maximum processing time with PEF is about 120 μs. This short processing time and low temperature (<40°C) are key factors that ensure no alteration of protein bands in the native PAGE electrophoretographs.

It has been reported that a more than seven log (>7 D) reduction of viable *E. coli* in LWE was achieved after 50 pulses of PEF, with an intensity of 48 kV/cm (Ma et al., 1997). Therefore, PEF processing with up to 50 pulses is recommended for future LWE pasteurization, without risk of protein coagulation.

CONCLUSIONS

The PEF process did not cause any changes in the electrophoretic patterns of LWE proteins, though some were observed with thermal and HHP processing. These changes were found to be a function of the applied pressure in the HHP process, or of the temperature in the thermal process. Both PEF and HHP demonstrated great potential as alternative or complementary processes to traditional methods of food processing. However, HHP was constrained by egg protein coagulation or denaturation at pressures of 60 kpsi or greater.

REFERENCES

Ball, H. R., Hamid-Samimi, M., Foegeding, P. M., and Swartzel, K. 1987. Functionality and microbial stability of ultrapasteurized aseptically packaged refrigerated whole egg. *J. Food Sci.* 52:1212–1218.

Banwart, G. J. 1989. *Basic Food Microbiology.* AVI Van Nostrand Reinhold, New York.

Brant, A. W., Patterson, G. W., and Walters, R. E. 1968. Batch pasteurization of liquid whole egg I. Bacteriological and functional properties evaluation. *Poult. Sci.* 47:878–885.

Chang, P. K., Powrie, E. D., and Fennema, O. 1970. Effect of heat treatment on viscosity of yolk. *J. Food Sci.* 35:864.

Cotterill, O. J., Glauert, J., and Bassett, H. J. 1976. Emulsifying properties of salted yolk after pasteurization. *Poult. Sci.* 55:544–548.

Cunningham, F. E. 1995. Egg product pasteurization. In: *Egg Science and Technology,* 4th edition. W. J. Stadelman and O. N. Cotterill, Eds. Food Product Press, Inc., New York.

Dixon, D. K. and Cotterill, O. J. 1981. Electrophoretic and chromatographic changes in egg-yolk proteins due to heat. *J. Food Sci.* 46:981.

Donovan, J. W., Mapes, C. J., Davis, J. G., and Garibaldi, J. A. 1975. A differential scanning calorimetric study of the stability of egg white to heat denaturation. *J. Sci. Food Agric.* 26:73.

Foegeding, P. M. and Stanley, N. W. 1987. Growth and inactivation of microorganisms isolated from ultrapasteurized egg. *J. Food Sci.* 52:1219.

Fujii, T., Satomi, M., Nakatsuka, G., Yamaguchi, T., and Okuzumi, M. 1994. Changes in freshness indexes and bacterial flora during storage of pressurized mackerel. *J. Food Hyg. Soc. Japan.* 35:195–200.

Grahl, T., Sitzmann, W., and Märkl, L. 1992. Killing of microorganisms in fluid media by high voltage pulses. *DECHEN4A Biotechnol. Conference Series,* 5B, pp. 675–678.

Hamid-Samimi, M. H. and Swartzel, K. R. 1985. Maximum changes in physical and quality parameters of fluid food during continuous flow heating. Applications to liquid whole egg. *J. Food Proc. Preserv.* 8:225–229.

Hegg, P. O., Martens, H., and Lofquist, B. 1978. The protective effect of sodium dodecylsulphate on the thermal precipitation of conalbumin. A study on thermal aggregation and denaturation. *J. Sci. Food. Agric.* 29:245.

Hite, B. H. 1989. The effects of pressure in the preservation of milk. *W. Va. Univ. Agric. Exp. Sta. Bull.* 58:15–35.

Ibarz, A., Sangronis, E., Ma, L., Barbosa-Cánovas, G. V., and Swanson, B. G. 1996. Viscoelastic properties of egg gels formed under high hydrostatic pressure. *IFT Annual Meeting.* New Orleans, LA. Abstract #80A-10.

Laemmli, U. K. 1970. Cleavage of structural proteins during assembly of the head of bacteriophage T4. *Nature.* 227:680–684.

Ma, L., Chang, F. J., Barbosa-Cánovas, G. V., and Swanson, B. G. 1997. Inactivation of *E. coli* in liquid whole egg using pulsed electric fields. *IFT Annual Meeting.* Orlando, FL.

Matsuda, T., Watanabe, K., and Sato, Y. 1981. Heat induced aggregation of egg white proteins as studied by vertical flat sheet polyacrylamide gel electrophoresis. *J. Food Sci.* 46:1829–1834.

Matsumoto, Y., Shioji, N., Imayasu, S., Kawato, A., and Mizuno, A. 1990. Sterilization of sake using pulsed high voltage application. *Proceedings of the 1990 Annual Meeting, Institute Electrostatics.* Japan. pp. 33–36.

Nakamura, R., Umemura, O., and Takemoto, H. 1979. Effect of heating on the functional properties of ovotransferrin. *Agric. Biol. Chem.* 43:325.

Potter, N. N. and Hotchkiss, J. H. 1995. Meat, poultry and egg. In: *Food Science,* 5th edition. Chapman and Hall, New York. Chapter 14, pp. 316–344.

Powrie, W. D. and Nakai, S. 1985. Characteristics of edible fluids of animal origin. Egg. In: *Food Chemistry.* O. R. Fennema, Ed. Marcel Dekker Inc., New York. Chapter 14, pp. 829–856.

Ronsivali, L. J. and Vieira, E. R. 1992. Poultry and egg. In: *Elementary Food Science.* AVI, Van Nostrand Reinhold, New York. Chapter 17, pp. 228–239.

Rovere, P. 1995. The third dimension of food technology. *Tech. Alimentari.* 4:1–8.

Sale, A. J. H. and Hamilton, W. A. 1967. Effects of high electric fields on microorganisms: 1. Killing of bacteria and yeast. *Biochim. Biophys. Acta.* 148:781–788.

Torten, J. and Eisenberg, B. 1982. Studies on colloidal properties of whole egg magma. *J. Food Sci.* 47:1423–1428.

Vanderzant, C. and Splittstoesser, F. D. (Eds.). 1992. *Compendium of Methods for the Microbiological Examination of Foods.* 3rd edition. American Public Health Association, Washington, DC.

Watanabe, K., Hayakawa, S., Matsuda, T., and Nakamura, R. 1986. Combined effect of pH and sodium chloride on the heat induced aggregation of whole egg proteins. *J. Food Sci.* 51:1112–1114.

Woodward, S. A. and Cotterill, O. J. 1983. Electrophoresis and chromatography of heat treated plain, sugared and salted whole egg. *J. Food Sci.* 48:501–506.

CHAPTER 16

Shelf Stability, Sensory Analysis, and Volatile Flavor Profile of Raw Apple Juice after Pulsed Electric Field, High Hydrostatic Pressure, or Heat Exchanger Processing

S. L. HARRISON
F. J. CHANG
T. BOYLSTON
G. V. BARBOSA-CÁNOVAS
B. G. SWANSON

ABSTRACT

F RESHLY pressed raw apple juice was obtained from a commercial processor. Clarification was performed by ultrafiltration, centrifugation, or plate filtration. Each of the clarified juices was further processed by pulsed electric field (PEF), high hydrostatic pressure (HHP), or heat exchanger (HE) methods. Processed PEF, HHP, and HE apple juice were stored at both 4°C and 23°C for one month. Viability of yeasts and molds, coliforms, aciduric bacteria, *Salmonella* spp., and total plate counts were determined. Microbial spoilage was not detected for heat exchanger and HHP processing, regardless of the temperature or length of storage. A maximum shelf life of one month was observed for PEF processed apple juice stored at 4°C. PEF processed apple juice spoiled during storage at 23°C. Sensory evaluation of attributes for raw apple juice were similar for PEF, HHP, and heat processed apple juice. Generally, apple juice volatile flavor compound concentrations were unchanged or decreased slightly after processing, and during 1 month storage.

241

INTRODUCTION

Inactivation of microorganisms in food systems by pulsed electric fields (PEF) and high hydrostatic pressure (HHP) was first studied by Sale and Hamilton (1967) and Hite (1899), respectively. Very little research is focused on the shelf life or sensory quality of foods processed by PEF. Matsumoto et al. (1990) reported that no change to the taste or fragrance of sake resulted from PEF processing. Sensory evaluations of milk and orange juice by Grahl et al. (1992) indicated that taste did not deteriorate significantly during PEF treatment. Food processing with HHP is on the verge of becoming a worldwide accepted food process. Japanese researchers have published several articles reporting that HHP treated food retains fresh taste and flavor. By producing the first food products available to consumers in April of 1990 (Rovere, 1995), the Japanese are considered the current world leaders in HHP food processing. Increasing the shelf life with HHP by a few days may be considered successful. By extending the shelf life of a food that spoils within days, greater distribution ranges can be achieved. An example is the processing of mackerel by HHP, which results in a 4-day extension in shelf life as judged by freshness indices and sensory evaluation (Fujii et al., 1994).

To successfully process raw apple juice by either PEF or HHP, the minimum refrigerated shelf life needed for distribution purposes is 1 month (Baranowski, 1996). Microbial load, sensory analysis, and flavor profile analysis may be utilized to define the shelf life of PEF and HHP processed apple juice. The characteristic fruity aroma of apples is a result of volatile flavor compounds, primarily esters, aldehydes, and alcohols (Flath et al., 1967; Willaert et al., 1983). Two of the principal aroma compounds in apples are butyl and hexyl acetate esters (Bartley et al., 1985). To determine the shelf life of PEF and HHP processed apple juice, the microbiological load, volatile flavor compound, and sensory characteristics need to be monitored initially and over time.

Ogawa et al. (1990) stated that HHP processing would be valuable in maintaining the natural flavor and taste of juices. It is the objective of the current research to extend the shelf life of raw apple juice with a pulsed electric field process and a high hydrostatic pressure process, without impacting the sensory quality of the manufactured apple juice.

MATERIALS AND METHODS

CLARIFICATION OF PRESSED APPLE JUICE

Apples were obtained, macerated, and pressed in the pilot plant of a local apple juice processing facility to yield raw apple juice. Raw apple juice was treated for 15 min at room temperature with a commercial pectinase solution, 0.6 mL Clarex 5x (Solvay, Elkhart, IN) per L apple juice, to hydrolyze pectin. Next, 200 Bloom, Type B gelatin (Atlantic, Whiteplains, NY) was added at 0.2 g/L to the apple juice. After an additional 15 min, 0.25 g/L bentonite (KWK American Colloid Company, Arlington Heights, IL) was added to precipitate protein. The apple juice was allowed to settle for 4 hr at 20°C prior to siphoning the upper "racked" portion to be used in the clarification step. The allotted time for each step in the pectin-gelatin-bentonite procedure was shorter than that typically used, because of imposed travel and time restrictions. Less than 60% of the juice portion was recovered because of lack of precipitation of apple particulates.

ULTRAFILTRATION

Clarification of raw apple juice was achieved with an ultrafiltration system (model HF. Lab 5, Romicon, Inc., Woburn, MA) following manufacturer recommendations for use and cleaning. The PM50 (Romicon, Inc., Woburn, MA) hollow fiber cartridge was utilized. Water flux was estimated at 10 L/min. The filtrate was collected in 5 gal food grade pails, and frozen at −40°C until used. As early as 1921, researchers demonstrated storage of apple juice for at least 2 years at −8°C without noticeable loss of flavor, aroma, or color (Cruess et al., 1921). Ultrafiltered apple juice was thawed at 10°C for three days prior to experimental use.

CENTRIFUGATION

Raw apple juice was manipulated in a 4°C cold room. Four aliquots of juice were placed in four 1 L centrifugation buckets. Centrifugation (International Equipment Co., Needham Hts., MA) was performed at 4°C, at a relative centrifugal force of 1300 g for 10 min. Centrifuged apple juice was collected in a 20 L carboy, maintained in a 10°C cold room until used for experimental purposes the next day.

PLATE FILTRATION

Approximately 1% 700 flow Kenite diatomaceous earth (World Minerals, Lompoc, CA) was added to siphoned apple juice to assist in the plate filter process. Filtration was accomplished with a shell-and-leaf pressure filtration system at 15 psi and 20°C. Filtered apple juice was collected in 5 gal food grade pails, and stored overnight at 10°C prior to use in experiments.

PROCESSING OF CLARIFIED APPLE JUICE

Heat Exchanger

The portion of the heat exchanger placed in the heated water bath consisted of coiled stainless steel tubing with an internal diameter of 5 mm, a wall thickness of 0.5 mm, and a length of 7 m. Raw apple juice was pumped through the heat exchanger at a rate of 500 mL/min. A water bath provided the heat necessary to increase the temperature of raw apple juice, with an initial temperature of 12°C, to obtain an exit temperature of 90°C. The processed apple juice was collected into 1 L Scholle bags (type II. MET. W/800R IRR, Scholle Co., Northlake, IL). In addition, one hundred 4 mil polyethylene pouches (Power Plastics Inc., Paterson, NJ) were filled with 25 mL of heated apple juice to monitor microbiological growth over time.

Pulse Electric Field

A pilot plant size pulser (Physics International, San Leandro, CA), in conjunction with an autoclaved sterile recirculating coaxial electrode chamber (Zhang et al., 1995), was utilized for PEF processing. Commercially sterile apple juice supplied by a local processor was used to prime the PEF chamber and tubing in a recirculating manner at a flow rate of 500 ml/min. Gas bubbles were removed from the treatment chamber after addition of commercially sterile apple juice, and prior to application of PEF to reduce the possibility of arcing. Prior to PEF processing, the flow of commercially sterile apple juice was stopped, the inlet tubing was removed from the commercially sterile apple juice source and placed in the raw apple juice, and the direction of flow was operated in the reverse direction for ~2 s to remove any bubbles that may have entered the inlet tubing. Flow of raw apple juice was started at 500 mL/min, and PEF processing was immediately initiated. The PEF operation was run at room temperature (~22°C). An electric field of 65 kV/cm, capacitance of 0.5 μF, and exponential decay pulses of 4 μs were utilized. The PEF processed raw apple juice received 6 pulses per unit volume. The PEF processed apple juice was poured into 1 L Scholle bags (type II. MET. W/800R IRR, Scholle Co., Northlake, IL). In addition, one hundred 4 mil polyethylene pouches (Power Plastics Inc., Paterson, NJ) were filled with 25 mL PEF processed juice to monitor microbiological growth over time.

HIGH HYDROSTATIC PRESSURE PROCESSING

One liter of raw apple juice was poured into each 1 L Scholle bag (type II. MET. W/800R IRR, Scholle Co., Northlake, IL). The Scholle bags were placed in polyethylene pouches (Power Plastics Inc., Paterson, NJ), and 100 mL of sterile distilled water was added to each pouch, followed by removing as much air as possible and heat sealing. One hundred 4 mil polyethylene pouches (Power Plastics Inc., Paterson, NJ) were filled with 25 mL of raw apple juice to monitor microbiological growth over time. A pilot plant scale hydrostatic pressure system (Engineered Pressure Systems Inc., Andover, MA), operated at room temperature ($\sim$22°C), was utilized to process the Scholle bags and polyethylene pouches of apple juice. The pressure come up time was 3.5 min, the holding time at 85 kpsi was 5 min, and the pressure release time was less than 15 s.

VIABILITY OF MICROORGANISMS

Microbial viability was determined for clarified apple juices prior to processing and immediately after processing. At 3 to 4 day intervals, apple juice from two 4 mil polyethylene pouches (Power Plastics Inc., Paterson, NJ) was diluted in 0.1% peptone, and plated in PDA to monitor microbial growth throughout the one month storage period. When researching fruit juice spoilage, Hatcher et al. (1992) suggest aerobic plate count, aciduric bacteria, yeast and molds, *Geotrichum* count, heat resistant molds, and direct microscope counts. Aerobic plate count, aciduric bacteria, and yeast and mold counts were determined for the current study. In addition, *Salmonella typhimurim*, which can survive up to 30 days in pH 3.6 apple juice (Goverd et al., 1979), was monitored. Coliforms were also monitored as a measure of sanitation. Manufacturer recommendations for media preparation were followed. The *Compendium of Methods for the Microbiological Examination of Foods* (Vanderzant and Splittstoesser, 1992) was used as guide for conducting microbial experiments.

The viability of the microorganisms was monitored by counting colony forming units (cfu) in agar plates. Apple juice was serially diluted with 0.1% sterile peptone (DIFCO 01118-01-8) solution and plated in the appropriate media. Agar plates were stored in an inverted position for selected times and temperatures to obtain viable counts. Serial dilutions for the viable counts were performed such that the number of cfu in the agar plates was between 25 and 300. The mean viable count was calculated from four plates.

Yeast and Mold Viability

The serially diluted apple juice was pour plated in DIFCO 0013-17-6 potato dextrose agar (PDA), acidified with 14 mL/L of 10% filter sterilized tartaric acid. PDA plates were incubated at 24°C for 72 hr.

Coliform Viability

The serially diluted apple juice was pour plated in DIFCO 0012-17-7 violet red bile (VRB) agar. After solidification of the VRB agar, additional VRB agar was added to form an overlay. VRB agar plates were incubated at 35°C for 24 hr.

Aciduric Bacteria Viability

The serially diluted apple juice was pour plated in DIFCO 0521-17-1 orange serum agar (OSA). OSA plates were incubated at 30°C for 48 hr.

Salmonella Spp. Detection

The presence of *Salmonella* spp. was detected by adding 25 mL of apple juice to 225 mL sterile tryptic soy broth (DIFCO 0370-17-3), and incubating for 24 hr at 35°C. One milliliter portions of the 250 mL apple juice/tryptic soy broth were added to 10 mL selenite cystine broth (DIFCO 0687-17-1) and tetrathionate broth base (DIFCO 0104-17-6), and incubated at 35°C for 24 hr. The selenite cystine broth culture was streak cultured on bismuth sulfite (DIFCO 0073-17-3), XLD (DIFCO 0788-17-9), and hektoen enteric (DIFCO 0853-17-9) agars. The tetrathionate broth culture was also streak cultured on bismuth sulfite, XLD, and hektoen enteric agars. The agar plates were incubated at 35°C for 24 hr. Colonies with characteristics of *Salmonella* were identified as positive *Salmonella* matches.

Total Plate Count

The serially diluted apple juice was pour plated in DIFCO 0479-17-3 plate count agar (PCA). PCA plates were incubated at 35°C for 48 hr.

STORAGE OF APPLE JUICE

Processed apple juice was stored at 23°C, and in a walk-in cooler at 4°C, for one month. A portion of the unprocessed apple juice was stored at −40°C, and used as the control for the one month sensory evaluation test.

VOLATILE FLAVOR PROFILE ANALYSIS

Purge-and-trap headspace analysis techniques were used for the isolation of volatile flavor compounds, as described by Boylston et al. (1994). Apple juice (200 mL) was placed in a 500 mL round bottom flask. Tetradecane (internal standard, 5 μg) was added to the flasks prior to extraction of volatile flavor compounds. The control, PEF, HHP, and HE apple juice samples were placed in a 40°C water bath, and purged with prepurified nitrogen (25 mL/min). A vacuum was applied to the system through an activated Tenax TA, 60/80 mesh trap (Alltech Associates, Deerfield, IL), 300 mg, packed in 9 mm OD borosilicate glass tubes for the collection of volatiles. The apple juice samples were purged for 5 hr. Tridecane (second internal standard, 1.25 μg) was added to the top of each trap and the volatiles were eluted from the traps with 15 mL hexane (HPLC grade), and concentrated to 200 μL under prepurified nitrogen. The tridecane was used to determine the percent of compound elution through the trap.

The volatile flavor compounds were separated on a 5% phenyl, 95% methylpolysiloxane capillary column (SE-54, 30 m, 0.32 mm ID, Alltech Associates, Deerfield, IL), installed in a gas chromatograph (Model 5880, Hewlett Packard, Avondale, PA) equipped with a flame ionization detector and on-column injection port. The program for the GC oven was 35°C for 15 min, increased to 160°C at 1.5°C/min, then increased to 225°C at 5°C/min and held for 3 min. The temperature was maintained at 250°C for both the injector and detector.

Area counts vs. nanogram quantities (6.25 to 200 ng) were plotted for the internal standard, tetradecane. Peak areas for detected volatile flavor compounds, based on retention time, were converted to nanogram quantities, based on the tetradecane standard curve. Contents of the individual volatile compounds were calculated, based on the recovery of the internal standard and the 200 mL quantity of apple juice used in the isolation. Identification of the volatile compounds in the isolates was based on comparison of GC retention times to pure commercial standards.

SENSORY ANALYSIS

Sensory analysis was performed on apple juice three days after processing to allow time to determine microbiological loads. Apple juice that remained unspoiled at the end of the 1 month storage period was used in sensory evaluation studies. Spoilage was defined as any objectionable flavors, odors, aftertaste, or $\geq 5 \times 10^3$ organisms/mL apple juice.

Panelists evaluated apple juice in individual booths under red lights for difference tests, and under white lights for acceptability tests. The apple juice was placed into 2 ounce clear plastic condiment cups fitted with lids. Low salt crackers and distilled water were provided to clean the palate between apple juice samples. The apple juice treatments were identified with a three digit number. The order of apple juice treatment presentation to the panelists was balanced and randomized.

Difference Test

Difference tests were performed three days and one month after storage of processed apple juice. Thirty-five untrained consumer panelists participated in each difference test. Triangle tests were used to evaluate the apple juice treatments.

Acceptability Test

Ninety-five panelists, 56 females and 39 males, with a mean age of 30, responded to a questionnaire and evaluated apple juice. Each panelist was asked if he/she consumed apple juice and how many servings each of fresh, frozen concentrate, and bottled apple juice were ingested per month. Eighty-seven panelists consumed apple juice at least once per month. The remaining eight panelists did not consume apple juice. Of the total 647 apple juice servings ingested per month by the 95 panelists, the number of fresh, frozen concentrate, and bottled servings consumed were 21%, 28%, and 51%, respectively. The panelists evaluated the processed apple juices for overall acceptability and for sweet, sour and color characteristics, using a nine point hedonic scale, with 9 = like extremely and 1 = dislike extremely. Each panelist evaluated all apple juice treatments. Panelists were presented with two plates with either two or three samples per plate. The order of apple juice presentation to the panelists was randomized.

Trained Panel Test

Eleven trained panelists from a commercial apple juice processing plant rated processed apple juice after one week and one month storage, using a rating scale of 1 (poor quality) to 5 (high quality).

STATISTICAL ANALYSIS

The critical number (minimum) of correct answers table in Meilgaard et al. (1991), was used to determine significance in the triangle difference tests. Significant differences between means for acceptability tests were established at $P \leq 0.05$, and determined using least square means (SAS, 1998).

RESULTS AND DISCUSSION

Sensory panelists participating in the difference analysis were able to detect statistical differences between at least one pair of apple juice treatments for each sensory taste panels performed (Table 16.1). Difference tests were used to determine if processing conditions contributed to significant differences between apple juice treatments that could be perceived by panelists. The nature of observed differences or specific sensory attributes altered by processing conditions is not elucidated by difference tests (Meilgaard et al., 1991). Perceived differences between apple juice processed by different means warranted further research to determine what characteristics might be altered by specific processes.

TABLE 16.1. Results of Sensory Triangle Differences Tests.

Experiment	Process vs.	Process	Significantly Different at
Ultrafiltered (time zero)	Heat exchanger vs.	PEF	$P \leq 0.05$
		HHP	ns
		Control	ns
Centrifuged (time zero)	Heat exchanger vs.	PEF	$P \leq 0.01$
		HHP	$P \leq 0.01$
		Control	$P \leq 0.01$
	Control vs.	PEF	ns
		HHP	ns
	PEF vs.	HHP	ns
Plate filtered (time zero)	Heat exchanger vs.	PEF	ns
		HHP	ns
		Control	$P \leq 0.1$
Plate filtered (1 month)	Control vs.	PEF (4°C)	ns
		HHP (4°C)	ns
		Heat exchanger (4°C)	$P \leq 0.1$
		PEF (23°C)	$P \leq 0.01$
		HHP (23°C)	$P \leq 0.05$
		Heat exchanger (23°C)	ns

ns = Not significantly different.

ULTRAFILTERED APPLE JUICE PROCESSING

Ultrafiltration (UF) of apple juice was adequate to extend the microbiological shelf life of the juice to 30 days (Table 16.2). However, a noticeable "off" flavor, associated with a strong aftertaste, was easily detected in the UF juice by the end of one month storage, regardless of microbiological loads. PEF

TABLE 16.2. Outgrowth of Microorganisms in Processed Ultrafiltered Apple Juice.

Temperature of Storage	Processing Method	Microorganism Outgrowth On	cfu/mL	Number of Days after Processing
4°C	None (control)	APDA	3×10^3	30
		OSA	3×10^3	
	PEF	None	nd	n/a
	HHP	None	nd	n/a
	Heat exchanger	None	nd	n/a
23°C	None (control)	OSA	1×10^6	16
	PEF	OSA	2×10^3	16
	HHP	None	nd	n/a
	Heat exchanger	None	nd	n/a

APDA = acidified potato dextrose agar; OSA = orange serum agar; nd = less than 25 cfu/mL detected; n/a = not applicable.

TABLE 16.3. Sensory Acceptability of Ultrafiltered Apple Juice Three Days after Processing.

Attribute	Heat Exchanger	PEF (4°C)	HHP (4°C)	Control (4°C)
Overall	5.54ab	5.38b	5.83ab	5.92a
Sweetness	5.84a	5.90a	6.22a	6.29a
Sourness	3.57a	3.54a	3.62a	3.53a
Fresh flavor	5.01a	4.91a	5.40a	5.43a
Color	7.30a	7.28a	7.44a	7.37a

Values in the same row with the same letter are not significantly different ($P \leq 0.05$).

processing at 65 kV/cm and 6 pulses resulted in a 4°C microbiological shelf-stable product, but an aciduric bacteria grew slowly during room temperature storage. Processing UF apple juice by HHP, at 85 kpsi for one 5 min cycle, achieved a 4°C and a 23°C microbiologically stable apple juice product. Most of the microorganisms were removed during ultrafiltration, but the enzymes were most likely not removed. Enzyme transmission through the UF membrane normally ranges from 50% to 90%, depending on the extent of membrane fouling (Gutman, 1987). In addition, PEF and HHP processing techniques do not completely inactivate enzymes (Castro, 1994; Vega, 1996).

During ultrafiltration, apple juice was constantly pumped through a filter cartridge system, and exited to a holding tank before being recycled. Oxygen is readily incorporated during the UF process. Fruit juices develop characteristic unpleasant off flavors when improperly pretreated, and when oxygen is not properly excluded (Hard, 1985). The specific chemical mechanisms of such flavor changes are not clear at this time, but can be largely avoided if oxidative enzymes are inactivated. Also, the duration of the filter operation could lead to losses in volatile flavor compounds through adsorption on the ultrafilter membrane surface, as well as to vaporization through recirculation of the retentate (Rao et al., 1987). Panelists found the characteristics of ultrafiltered juice from the different processing methods to be equivalent (Table 16.3), with the exception of "overall" acceptability.

CENTRIFUGED APPLE JUICE PROCESSING

Centrifugation (CF), in itself, was not adequate to reduce the microbial population to acceptable levels (Table 16.4). The refrigerated shelf life of centrifuged apple juice without further processing was less than one week, as noted by the 4°C and 23°C control apple juice in Table 16.4. Additional processing beyond CF by PEF, HHP, or heat exchanger resulted in a microbiologically stable refrigerated apple juice for one month. PEF processed CF apple juice

TABLE 16.4. Outgrowth of Microorganisms in Processed Centrifuged Apple Juice.

Temperature of Storage	Processing Method	Microorganism Outgrowth On	cfu/mL	Number of Days after Processing
4°C	None (control)	PCA	7×10^4	4
		APDA	1×10^7	9
		OSA	1×10^6	9
		VRB	4×10^3	9
	PEF	None	nd	n/a
	HHP	None	nd	n/a
	Heat exchanger	None	nd	n/a
23°C	None (control)	PCA	5×10^3	4
		APDA	3×10^5	4
		OSA	4×10^6	4
	PEF	APDA	3×10^6	9
		OSA	3×10^6	9
	HHP	None	nd	n/a
	Heat exchanger	None	nd	n/a

PCA = plate count agar; APDA = acidified potato dextrose agar; OSA = orange serum agar; VRB = violet red bile agar; nd = less than 25 cfu/mL detected; n/a = not applicable.

stored at room temperature was spoiled by both yeast and bacteria within nine days of processing. HHP and heat exchanger processed CF apple juice resulted in no observable microbiological growth at either 4°C or 23°C for one month.

The sensory scores for the acceptability of centrifuged apple juice (Table 16.5) were approximately one point higher, across the table, as compared to the ratings by panelists for the UF processed apple juice (Table 16.3), indicating that CF apple juice is more acceptable than UF apple juice. Only the "overall" acceptability attribute was determined to be significantly different between two of the processing methods, PEF and heat exchanger.

TABLE 16.5. Sensory Acceptability of Centrifuged Apple Juice Three Days after Processing.

Attribute	Heat Exchanger (23°C)	PEF (4°C)	HHP (4°C)	Control (4°C)
Overall	6.97a	6.27b	6.72ab	6.61ab
Aroma	6.82a	6.61a	6.51a	6.33a
Sweet	6.76a	6.73a	6.71a	6.94a
Sour	3.70a	3.58a	3.54a	3.54a
Fresh flavor	6.77a	6.35a	6.57a	6.53a
Color	5.98a	5.84a	6.25a	5.82a

Values in the same row with the same letter are not significantly different ($P \leq 0.05$).

TABLE 16.6. Outgrowth of Microorganisms in Processed Plate Filtered Apple Juice.

Temperature of Storage	Processing Method	Microorganism Outgrowth On	cfu/mL	Number of Days after Processing
4°C	None (control)	PCA	2×10^4	0
		APDA	2×10^3	0
		OSA	1×10^4	0
	PEF	PCA	8×10^1	29
		APDA	3×10^2	29
		OSA	7×10^1	29
	HHP	None	nd	n/a
	Heat exchanger	None	nd	n/a
23°C	None (control)	PCA	2×10^4	0
		APDA	2×10^3	0
		OSA	1×10^4	0
	PEF	PCA	3×10^6	3
		APDA	3×10^6	3
		OSA	3×10^6	3
	HHP	None	nd	n/a
	Heat exchanger	None	nd	n/a

PCA = plate count agar; APDA = acidified potato dextrose agar; OSA = orange serum agar; VRB = violet red bile agar; nd = less than 25 cfu/mL detected; n/a = not applicable.

PLATE FILTERED APPLE JUICE PROCESSING

The initial microbiological load for the plate filtered (PF) apple juice was much larger than for either the UF (Table 16.2) or CF (Table 16.4) processed juice (Table 16.6). Even with high initial loads, PEF processed, plate filtered juice obtained an extended shelf life at refrigeration temperatures. Whereas, PEF treated apple juice at 23°C spoiled rapidly, and HHP and heat exchanger processed apple juice maintained microbial shelf stability at 4°C and 23°C for the one month study.

The sensory data collected for the plate filtered (PF) acceptability panel (Table 16.7) was quite divergent from the previous ultrafiltered (Table 16.3) and centrifuged (Table 16.5) acceptability panels. Three of the six apple juice attributes were significantly different for the PF apple juice treatments (Table 16.7). Surprisingly, the "overall" acceptability attribute was not significantly different for any pair of apple juice processing methods for the PF apple juice. The three day acceptability sensory scores for the "color" attribute were significantly lower for the heated apple juice than for the apple juice processed by PEF, HHP, or the control apple juice. The heated apple juice was slightly darker, which was considered a negative attribute. The overall mean sensory scores remained high for plate filtered juice, being similar to CF sensory scores (Table 16.4), and about 1 point higher than sensory scores from the UF experiment (Table 16.2).

TABLE 16.7. Sensory Acceptability of Plate Filtered Apple Juice Three Days after Processing.

Attribute	Heat Exchanger (23°C)	PEF (4°C)	HHP (4°C)	Control (4°C)
Overall	6.42a	6.85a	6.85a	6.46a
Aroma	6.59ab	6.95a	6.71ab	6.39b
Sweetness	6.99ab	7.28a	7.09ab	6.74b
Sourness	3.41a	3.17a	3.33a	3.30a
Fresh flavor	6.40a	6.54a	6.58a	6.16a
Color	7.10b	7.66a	7.61a	7.37ab

Values in the same row with the same letter are not significantly different ($P \leq 0.05$).

Plate filtered apple juice from PEF, HHP, or heat exchanger processing methods and storage condition (4°C or 23°C) had low microbial counts, and the apple juice passed an informal tasting session. An exception was the PEF treated apple juice stored at 23°C, which spoiled after just three days at room temperature (Table 16.6) and, therefore, was not tasted in the informal tasting session. The remaining apple juice tasted was palatable. The portion of the control frozen at −40°C, to be used as the control for the one month sensory tests, was thawed for three days in a 10°C cold room prior to the one month sensory tests.

According to Cruess et al. (1921), apple juice can be stored at −8°C for up to two years without changes to flavor, aroma, or color. The apple juice samples processed by PEF and HHP were expected to, score equal to, or better than, the frozen control and the commercially available apple juice. Significant differences were observed for every attribute studied after one month of storage (Table 16.8). The commercial apple juice obtained the lowest sensory scores for each attribute except sourness. The HHP (4°C) sensory scores were equal to the control for every attribute, and greater than the commercial apple juice sensory scores, except for the "sourness" attribute. Sensory scores for PEF (4°C), HHP

TABLE 16.8. Sensory Acceptability of Plate Filtered Apple Juice One Month After Processing.

Attribute	Commercial (23°C)	Heat Exchanger (23°C)	PEF (4°C)	HHP (4°C)	HHP (23°C)	Control (4°C)
Overall	4.14c	6.11ab	5.96ab	6.21a	5.62b	6.35a
Aroma	4.75c	6.22ab	5.69b	6.10ab	6.22ab	6.26a
Sweetness	4.71b	6.72a	6.31a	6.55a	6.72a	6.61a
Sourness	4.63a	2.63b	2.67b	3.05a	2.82b	3.09b
Fresh flavor	4.02c	5.93ab	5.65ab	5.96ab	5.39b	6.05a
Color	6.20b	6.93a	6.97a	7.34a	7.00a	7.14a

Values in the same row with the same letter are not significantly different ($P \leq 0.05$).

TABLE 16.9. Trained Panel Sensory Scores for Plate Filtered Apple Juice One Month After Processing.

Treatment	Sensory Score
Control (4°C)	1.93b
Heat exchanger (4°C)	3.45a
Heat exchanger (23°C)	3.21a
PEF (4°C)	3.38a
HHP (4°C)	3.53a
HHP (23°C)	2.85a

Values with the same letter are not significantly different ($P \leq 0.05$).

(23°C), and heat exchanger (23°C) apple juice were generally higher than the commercial apple juice sensory scores, but lower than the control apple juice sensory scores. The trained panel rated the control apple juice very poorly with the remainder of the apple juice treatments rated as equal (Table 16.9). The control apple juice used for the trained panel was stored at 4°C for one week prior to sensory evaluation, resulting in a partially fermented product as described by the panelists.

The commercial apple juice and apple juice used in the remaining treatments were produced from different lots of apples, and, therefore, are not directly comparable to each other. The commercial apple juice and apple juice used in the remaining treatments were produced from apples stored for the same length of time, but most likely came from two different orchards. Because of the immense scale of the commercial operation, a direct comparison between commercial, PEF, HHP, and heated apple juice was not possible.

It was anticipated that analysis of the volatile flavors in apple juice processed by these innovative methods (PEF and HHP) would yield some insight into the chemical changes taking place that affect sensory scores. The volatile flavor analysis of five selected compounds of importance to apple juice flavor is listed in Table 16.10, with respect to the three experiments involving ultrafiltered, centrifuged, or plate filtered apple juice. Initial butyl and hexyl hexanoate concentrations were greater in the clarified UF and plate filtered control apple juice, and decreased after further processing. The centrifuged control apple juice butyl and hexyl hexanoate concentrations were equal to, or lower than, the apple juice with further processing. Hexyl acetate was markedly lower for both ultrafiltration and centrifugation clarified apple juice as compared to plate filtered apple juice controls. The remainder of the volatile flavor data is otherwise similar, and does not demonstrate striking differences between experiments or within experiments. Boylston et al. (1994), employing a similar volatile flavor extraction method for Gala apples stored for four months, reported hexanal, butyl acetate, hexyl acetate, butyl hexanoate, and hexyl hexanoate concentrations of 22.8, 236.7, 224.0, 39.4, and 39.6 ng/mL,

TABLE 16.10. Contents* of Volatile Flavor Compounds from Apple Juice, as Influenced by Processing and Storage Time.

Experiment	Treatment	Hexanal	Butyl Acetate	Hexyl Acetate	Butyl Hexanoate	Hexyl Hexanoate
Ultrafiltration	Control	3.0b	74.4a	5.2bc	0.9a	4.9a
(time zero)	PEF	9.7a	179.1a	19.0ab	0.3a	0.4b
	HHP	7.7a	89.4a	32.7a	0.1a	0.3b
	HE	6.4ab	81.1a	2.4c	0.7a	0.1b
Centrifugation	Control	4.1a	15.7b	14.3ab	0.2b	0.2a
(time zero)	PEF	1.5a	8.9b	9.6b	0.9b	0.5a
	HHP	3.1a	3.3b	11.6ab	3.6a	0.5a
	HE	0.3a	11.6b	10.3b	0.5b	0.5a
	Commercial	23.4a	66.1a	20.2a	0.9b	0.1a
Plate	Control	4.1aX	58.3aX	53.2aX	3.8aX	9.3aX
filtration	PEF	3.0bcX	38.8bY	41.9bY	0.2bX	0.8bX
(time zero)	HHP	2.5cX	26.5cX	41.0bX	0.0bX	0.0bX
	HE	3.8abX	40.3bX	36.3bX	0.1bX	0.2bX
Plate	Control	3.3aX	37.8bcY	43.6abY	1.0aY	1.6aY
filtration	PEF 4°C	3.7aX	50.3aX	51.3aX	0.5abX	0.6abX
(one month)	HHP 4°C	2.2aX	28.8cX	38.8bcY	0.0bX	0.1bX
	HHP 23°C	2.5aX	25.1cX	36.4bcY	0.0bX	0.1bX
	HE 23°C	3.6aX	35.8bcX	35.1cX	0.1bX	0.1bX
	Commercial	4.1a	39.9ab	19.9d	0.9a	0.7ab

*1 ng/mL apple juice.
Values for each experiment contained in the same column with the same lower case letter are not significantly different ($P \leq 0.05$).
Changes in volatile compound concentrations due to storage time for the plate filtration experiment are indicated by the upper case letters X and Y. If the time zero and one month values are not significantly different, the values will be followed by the same upper case letter.
PEF = pulsed electric field; HHP = high hydrostatic pressure; HE = heat exchanger; control = clarified with no further processing; and commercial = shelf stable commercially processed apple juice.

respectively. The apples utilized by Boylston et al. (1994) were intended for the fresh market, whereas apples in the current study were intended for apple juice production.

CONCLUSIONS

Raw apple juice can be treated by either pulse electric field (PEF) or high hydrostatic pressure (HHP) methods to extend the shelf life of apple juice to one month. In addition to a prolonged shelf life, the acceptability of the PEF and HHP processed apple juice stored at 4°C is maintained equal to that of fresh pressed apple juice, and is more acceptable to consumers than the commercial apple juice tested. Even longer apple juice shelf life extensions may be possible with: (1) PEF, if a combination of more intense electric fields or more pulses are used; or (2) HHP, if a combination of higher pressures and longer times are utilized. Overall, the differences observed with GC suggest that few, if any, changes are occurring to the volatile chemicals in apple juice due to PEF or HHP processing.

ACKNOWLEDGEMENTS

The authors wish to thank Humberto Vega-Mercado and Frank Younce for their assistance with the operation of the WSU pulser. The funding for this research was provided by the U.S. Army Natick Research Development and Engineering Center, Natick, MA, and the Bonneville Power Administration, Department of Energy, Walla Walla, WA.

REFERENCES

Baranowski, J. 1996. Personal communication. Tree Top Inc., Selah, WA.

Bartley, I. M., Stoker, P. G., Martin, A. D. E., Hatfield, S. G. S., and Knee, M. 1985. Synthesis of aroma compounds by apples supplied with alcohols and methyl esters of fatty acids. *J. Sci. Food Agric.* 36:567–574.

Boylston, T. D., Kupferman, E. M., Foss, J. D., and Buering, C. 1994. Sensory quality of Gala apples as influenced by controlled and regular atmospheric storage. *J. Food Qual.* 17:477–494.

Castro, A. J. 1994. Pulsed electric fields modification of activity and denaturation of alkaline phosphatase. Ph.D. dissertation, Wash. State Univ., Pullman, WA.

Cruess, W. V., Overholser, E. L., and Bjarnason, S. A. 1921. The storage of perishable fruits in freezing storage. *Univ. Calif. Exp. Sta. Bull. No. 241.*

Flath, R. A., Black, D. R., Guadagni, D. G., McFadden, W. H., and Schultz, T. H. 1967. Identification and organoleptic evaluation of compounds in Delicious apple essence. *J. Agric. Food Chem.* 15:29–35.

Fujii, T., Satomi, M., Nakatsuka, G., Yamaguchi, T., and Okuzumi, M. 1994. Changes in freshness indexes and bacterial flora during storage of pressurized mackerel. *J. Food Hyg. Soc. Japan.* 35:195–200.

Goverd, K. A., Beech, F. W., Hobbs, R. P., and Shannon, R. 1979. The occurrence and survival of coliforms and salmonellae in apple juice and cider. *J. Appl. Bacteriol.* 46:521–530.

Grahl, T., Sitzmann, W., and Märkl, H. 1992. Killing of microorganisms in fluid media by high-voltage pulses. *DECHEMA Biotechnology Conference Series,* 5B, 675–678.

Gutman, R. G. 1987. Applications. Chapter 5 in *Membrane Filtration: The Technology of Pressure-Driven Crossflow Processes.* pp. 130–166. Adam Hilger, Bristol, England.

Hard, N. F. 1985. Characteristics of edible plant tissues. Chapter 15 in *Food Chemistry,* 2nd ed. O. R. Fennema (Ed.), pp. 857–912. Marcel Dekker, Inc., New York.

Hatcher, W. S. Jr., Weihe, J. L., Splittstoesser, D. F., Hill, E. C., and Parish, M. E. 1992. Fruit beverages. Chapter 51 in *Compendium of Methods for the Microbiological Examination of Foods.* C. Vanderzant and D. F. Splittstoesser (Eds.), pp. 953–974. American Public Health Association, Washington, DC.

Hite, B. H. 1899. The effects of pressure in the preservation of milk. *W. Va. Univ. Agric. Exp. Sta. Bull.* 58:15–35.

Matsumoto, Y., Shioji, N., Imayasu, S., Kawato, A., and Mizuno, A. 1990. Sterilization of "Sake" using pulsed high voltage application. *Proceedings of 1990 Annual Meeting Institute Electrostatics,* Japan, pp. 33–36.

Meilgaard, M. M., Civille, G. V., and Carr, B. T. (Eds.). 1991. *Sensory Evaluation Techniques,* 2nd ed. CRC Press, Inc., Boston, MA.

Ogawa, H., Fukuhisa, K., Kubo, Y., and Fukumoto, H. 1990. Pressure inactivation of yeasts, molds, and pectinesterase in Satsuma Mandarin juice: Effects of juice concentration, pH, and organic acids, and comparison with heat sanitation. *Agric. Biol. Chem.* 54:1219–1225.

Rao, M. A., Acree, T. E., Cooley, H. J., and Ennis, R. W. 1987. Clarification of apple juice by hollow fiber ultrafiltration; fluxes and retention of odor-active volatiles. *J. Food Sci.* 52:375–377.

Rovere, P. 1995. The third dimension of food technology. *Tech. Alimentari.* 4:1–8.

Sale, A. J. H. and Hamilton, W. A. 1967. Effects of high electric fields on microorganisms: I. Killing of bacteria and yeast. *Biochim. Biophys. Acta.* 148:781–788.

SAS. 1998. *SAS User's Guide.* Statistical Analysis Systems Institute, Cary, NC.

Vanderzant, C. and Splittstoesser, F. D. (Eds.). 1992. *Compendium of Methods for the Microbiological Examination of Foods,* 3rd ed. American Public Health Association, Washington, DC.

Vega, H. 1996. Inactivation of proteolytic enzymes and selected microorganisms in foods using pulsed electric fields. Ph.D. dissertation, Wash. State Univ., Pullman, WA.

Willaert, G. A., Dirinck, P. J., De Pooter, H. L., and Schamp, N. N. 1983. Objective measurement of aroma quality of Golden Delicious apples as a function of controlled-atmosphere storage time. *J. Agric. Food Chem.* 31:809–813.

Zhang, Q., Barbosa-Cánovas, G. V., and Swanson, B. G. 1995. Engineering aspects of pulsed electric field pasteurization. *J. Food Eng.* 25:261–281.

CHAPTER 17

Pulsed Electric Field Treatment of Food and Product Safety Assurance

H. L. M. LELIEVELD
P. C. WOUTERS
A. E. LEÓN

P ULSED electric field (PEF) treatment of food may be applied for various reasons. One is the inactivation of microorganisms at temperatures below those significantly adversely affecting product characteristics such as color and flavor. Before applying any new technology in practice, there must be clear evidence that after processing the product is safe. Consequently, work is needed to prove that relevant food poisoning microorganisms are inactivated. If PEF is intended to replace pasteurization, it must be realized that although both PEF and heat treatments may inactivate microbial cells, the mechanisms of inactivation are likely to differ. While traditional pasteurization will also reduce the activity of many enzymes to a large extent, PEF is unlikely to have a similar effect. It must be shown that the lack of inactivation of certain enzymes does not have undesirable consequences for the safety of the product.

PEF is also likely to cause electrochemical changes, but these may be so small that product safety is not affected. Again, evidence is needed to verify this assumption. Some results of microbiological and toxicological investigations will be presented as well.

It is claimed that pulsed electric fields (PEF) inactivate microorganisms without affecting the quality of the food. If bacteria are changed to the extent that they die, why would nothing happen to the product itself?

Before applying PEF to foods, we have to *know*, and not merely assume, whether the final product is microbiologically and chemically safe. Although we know that PEF can destroy *Staphylococci, Yersinia*, and many other pathogenic and spoilage microorganisms, we also know from our own experiments, as well as from the literature, that there is a wide spread in the PEF sensitivity of microorganisms. The environment of the microorganisms also plays an important role.

259

For example, pH has a very significant effect: the lower the pH, the stronger the effect (Wouters et al., submitted). The number of microorganisms investigated so far is limited, and there might be PEF resistant ones. Even if that is not the case, whether microorganisms can develop such resistance by adapting to electric pulses should be investigated. There might be survivors with a different, more electrically resistive cell membrane, or a cell membrane that is physically stronger and does not fall apart so easily.

As explained by other researchers, cells may be porated by PEF. Originally, PEF was used to do just that. Porated cells may exchange DNA, and properties may be exchanged between microorganisms. It needs to be clear that the exchange of DNA or RNA between biological cells cannot result in adverse effects.

Damage to the microbial cell membrane may not necessarily be the most important cause of inactivation of the microorganism by PEF. Our own results have shown that under similar conditions, gram-negative microorganisms can be as resistant to PEF as gram-positive microorganisms. If PEF treated microorganisms are observed by microscope, one can see cells with a visibly destructed cell membrane, but most cells seem intact while not being viable any longer. Other causes of inactivation may be changes in transport proteins embedded in the membrane or damage to nucleic acid. More likely, however, all possible damage happen simultaneously, to various extents, depending on the immediate environment of the cell. These changes do not occur selectively to the microbial cell, but also potentially to eukaryotic cells. Destruction of the wall of both prokaryotic and eukaryotic cells may result in the release of enzymes in the product that otherwise would not be released. The enzyme activity may subsequently change the product.

Foods and food ingredients treated by a process not currently used in food production fall within the scope of the European Regulation on Novel Foods, if the process gives rise to significant changes in the composition, the structure, or the level of undesirable substances, and these changes affect the nutritional value or metabolism. A comparison has to be made between products processed conventionally and products produced with the novel process. If analyses show differences, then the influence on nutrition and metabolism has to be addressed. If the changes brought about by the novel process are significant (toxicologically and/or nutritionally), then a risk assessment must result in a decision on the safety status of the food. This decision will have to be endorsed by regulatory bodies.

If cells, that by traditional processing remain intact, burst as a result of PEF treatment, are the released cellular components safe for the consumer? This needs careful investigation. As is known, it takes just 30 micrograms of purified peanut protein to cause the death of a sensitized adult. The potential risk for anaphylactic shock due to allergy may be estimated by calculating how much otherwise contained microbial protein may be released into food. If a product contains a million bacteria per gram, it is noticeably spoiled, and one would not eat it. The dry weight of a million bacteria is approximately 0.1 μg. Assuming that 50% of the dry weight is protein, if all protein were released into the product it would contain 0.05 μg of bacterial protein per gram. A 100 gram portion then would contain 5 μg bacterial protein mixture. Although this does not prove the safety of a product containing such an amount of bacterial protein, the amount is significantly lower than the potentially deadly quantity of peanut protein.

Applying electrical pulses requires electrodes in contact with the food. If the electrodes are made from metal, corrosion may result in contamination of the food. Metals are acceptable in food, often desirable, but there are safety limits. The maximum concentration of heavy metals in food in many countries is 0.1 mg per kg. Again, the potential for serious risk may be calculated. As the speed of corrosion will strongly depend on the composition of the product in contact with the electrode, the rate of corrosion cannot be predicted. The dissolution of material from the electrodes, however, can be observed and easily quantified by determining the changes in dimension, or by weighing. For example, consider an extreme case: a PEF process unit producing 1000 kg per hr produces in three months time (91 days of 8 hr), 728,000 kg of product. If in that time an electrode with a diameter of 50 mm and a length of 10 mm loses 1 mm of its radius, the volume loss is 1600 mm^3. At a density of 9000 kg per m^3 (valid for copper and nickel), this equals 14,400 mg. Thus, the concentration in the product would be 144,000 mg metal/728,000 kg product, or ~0.02 mg per kg, far below the acceptable concentration. This amount would never seriously contribute to the required daily intake of iron, for example. It is unlikely that the corrosion rate would be as high as assumed, but a requirement could be to replace an electrode after a certain period of usage.

It has been suggested that the chance that products are changed electrochemically is very small, and that the only thing that may happen is the release of minute amounts of hydrogen and oxygen (nanograms per m^3 product) on the electrodes.

Pure-Pulse uses a system where a high voltage pulse is followed by a low voltage pulse of reversed polarity, such that the two charges are equal and the electrochemical effects are nullified. Whether relevant electrochemical reactions are slow enough for such a system to work should be investigated.

Others have suggested that the effect of PEF on microorganisms is at least partly the result of a secondary process, in particular the production of chlorine compounds. This could be so, as almost all food contains sodium chloride. Hülsheger and Niemann (1980) found some effect with *E. coli* (Figure 17.1 and 17.2). If starting with 10^6 *E. coli* cells, there is a higher inactivation rate in the presence of *Clostridium*. Starting with 10^8 cells, however, no effect is shown. Possibly, the amount of hypochlorite produced reacts with organic material that is present more abundantly when 100 times as many microorganisms are present. Experiments of Wouters (personal communication) did not show any influence of chloride ions on the inactivation rate of *Listeria* (Figure 17.3). A possible explanation of the difference in findings may be that Wouters used 10 times shorter pulses.

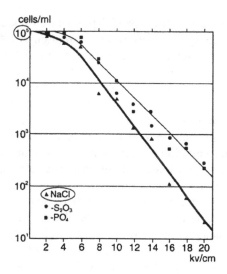

Figure 17.1 Relation between surviving cells and electric field for different electrolyte suspensions with 10^5 cells/mL (Hülsheger and Niemann, 1980).

Because of the legal requirements, we have asked RIKILT-DLO, a Dutch governmental food research institute, to investigate whether there are differences between PEF-treated and heat-treated tomato puree. RIKILT used high resolution one- and two-dimensional proton NMR fingerprinting. Some of the results are summarized in Table 17.1. The main differences were in the concentration

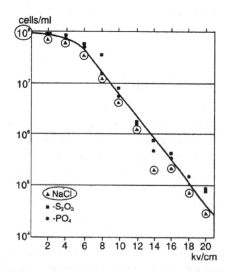

Figure 17.2 Relation between surviving cells and electric field for different electrolyte suspensions with 10^8 cells/mL (Hülsheger and Niemann, 1980).

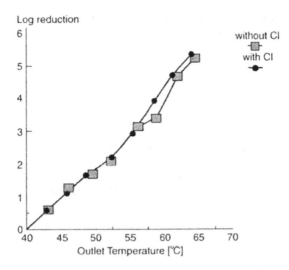

Figure 17.3 Effect of chlorine ions on inactivation of *Listeria innocua* NCTC 11289 in phosphate buffer (pH 5; conductivity 7.8 mS/cm at 40°C inlet temperature; field strength 2.5V/μm) (Wouter et al.).

of amino acids. This is not surprising, given that PEF does not inactivate enzymes, including proteases. The differences in sugar-like compounds are not as easy to explain. In total, RIKILT studied about 2000 peaks, and about 600 of those were different. A few hundred were significantly different. The question

TABLE 17.1. Comparison between Some Compounds in Tomato Purees Treated with Heat or with PEF (Based on High Resolution One- and Two-Dimensional Proton-NMR Fingerprinting by RIKILT-DLO, Wageningen, The Netherlands).

Compound	Concentration in PEF Treated Puree*
Water soluble compounds	
Glucose	−
Oligosaccharides	− −
Tyrosine	+
Valine, leucine, isoleucine	+ +
Sugar-like compounds	+ +
Fat soluble compounds	
Aromatic compounds	−
Sugar-like compounds	−/− −
Ribose-like compounds	+ +
Non-tomatine glycoalkaloids	+
Tomatine glycoalkaloids	− −
Indole type compounds	−

*Concentration in PEF treated product: − slightly lower, − − much lower, + slightly higher, + + much higher.

to be addressed is whether the observed differences are significant from the toxicological and/or nutritional point of view.

In conclusion, PEF potentially is an attractive process, but before application much research is still needed.

REFERENCES

Hülsheger, H. and Niemann, E. G. 1980. Lethal effects of high voltage pulses on *E. coli* k12. *Radiat. Environ. Biophys.* 18:281–288.

Wouters, P. C., Dutreux, N., Smelt, J. P. P. M., and Lelieveld, H. L. M. Effects of pulsed electric fields on inactivation kinetics of *Listeria innocua*. Appl. Env. Microb. 65(12):5364–5371.

Index

Milton Keynes UK
Ingram Content Group UK Ltd.
UKHW020024071024
449327UK00032B/2911